EMB-312 TUCANO, Brazil's turboprop success story

João Paulo Zeitoun Moralez

EMB-312 TUCANO

Brazil's turboprop success story

João Paulo Zeitoun Moralez

HARPIA PUBLISHING

2803 Sackett Street, Houston, TX 77098-1125, U.S.A.
enquiries@harpia-publishing.com

Consulting and inspiration by Kerstin Berger

Front cover artwork by Murilo Camargo Martins shows an FAB EMB-312 Tucano from 2ª Esquadrilha de Ligação e Observação (2ª ELO, Liaison and Observation Flight)

Rear cover artworks by Murilo Camargo Martins; from top to bottom: EMB-312 Tucanos of the FAB Esquadrilha da Fumaça and Mauretania, and a Kuwait Air Force Short Tucano T.Mk 52

Artworks by Murilo Camargo Martins

Editorial by Thomas Newdick

Layout by Norbert Novak, www.media-n.at, Vienna

Printed at finidr, Czech Republic

Harpia Publishing, L.L.C. is a member of

ISBN 978-0-9973092-3-2

Contents

Introduction

Until the 1980s, many air forces around the world utilised jets for the advanced training of their cadets or for the operational conversion of their future fighter and attack pilots. These training aircraft were highly manoeuvrable and had plenty of power for aerobatics and manoeuvres involving high G loads, but at the same time they were more costly to acquire, harder to maintain and much more expensive to operate due to their generally high fuel consumption.

Among the most popular aircraft in this class were the Aero L-29 Delfin, British Aircraft Corporation (BAC) Jet Provost, Cessna T-37, Fouga Magister and Morane-Saulnier MS.760 Paris. In general, these types reigned supreme for the high-performance training of military aviators.

But then a small Brazilian aeronautical manufacturer, Embraer, arrived with a new proposal for this military aviation sector. Even in its design phase, the new aircraft had to overcome many obstacles and challenge existing concepts to establish what would become new standards for advanced training aircraft.

In the late 1970s Brazil suffered difficulties obtaining spare parts for its fleet of 65 Cessna T-37Cs after the government broke a military assistance agreement with the United States. Recognising a requirement for the Brazilian Air Force, Embraer proposed the development of a new advanced trainer for local use. It would be 100 per cent indigenous and focused to meet Brazilian Air Force demands. However, Embraer already foresaw the sale of the aircraft on the foreign marketplace.

After fulfilling the first, basic steps of flying in simpler aircraft types, it was generally believed that the best arrangement was to have the aircraft's seats positioned side by side for the next phase of training. In this way, the instructor could better see the student's mistakes and correct them, while the cadet could replicate the instructor's control inputs to overcome any difficulties encountered during the flight.

But Embraer's proposal for what became the EMB-312 Tucano was a tandem-seat aircraft, with a single-piece canopy and ejection seats. The turboprop engine was aligned at a deflection of 3° from the longitudinal axis of the airframe. In addition, the cockpit would have a single lever instead of three – to control power, and the engine would be activated by simply pressing a button.

The arrangement of the instruments on their panels and the architecture of the cockpit allied to the other advances meant the Tucano emerged as a turboprop aircraft with the 'soul' of a jet. Its performance, manoeuvrability, speed and flexibility, as well as the internal architecture of the cockpit, which resembled that of a third-generation fighter, allowed higher-quality pilots to be trained at a much lower cost. The availability of the aircraft on the flight line was much improved and the transition to more advanced models could be made more quickly. Instruction in the Tucano was 20 per cent more efficient when compared to jet trainers in its category. Students needed 20 per cent fewer flight hours to assimilate the entire training programme, in an aircraft that consumed one-third of the the fuel of an equivalent jet aircraft.

But the Tucano also had another difference. From its inception it was expected to carry up to a tonne of weapons and stores on four underwing stations. These included unguided rockets, bombs and heavy machine guns.

In the operational training syllabus the Tucano served as a weapons trainer. Because of this capability, several air forces also used the Tucano as an armed escort aircraft, for close air support, reconnaissance and even the interception of small aircraft.

Since the Tucano entered service in 1983, aircraft flown by Honduras and Peru, for example, have been responsible for shooting down around 60 light aircraft used by drug traffickers on illegal and unauthorised flights. Colombian examples have also shot down several aircraft. In Angola, Colombia, Mauritania and Venezuela the Tucano has played a decisive role in various conflicts. In Brazil, it has helped to maintain sovereignty along the borders, tackling guerrillas and drug traffickers. In Paraguay, it has provided the only means of air defence for many years.

In total, 16 countries – including Brazil, have operated the Tucano. Of these, two are Western European countries, key NATO members and with permanent seats on the United Nations Security Council – France and the United Kingdom. In the UK, the Tucano was extensively modified and manufactured under licence by Short Brothers to meet Royal Air Force requirements and was also exported to Kenya and Kuwait.

In the US it has gained a new lease of life in the hands of enthusiasts and civilian companies dedicated to providing advanced military training for the world's air forces. In this capacity, new avionics and guided weapons have been integrated.

Embraer learned a lot from the Tucano. It learned how to develop a trainer, an aircraft with ejection seats, and the company pioneered new engineering and production techniques and refinements that had not previously been seen in its products.

The success of the EMB-312 in combat and patrol missions led to a new-generation aircraft – the EMB-314 Super Tucano. This has avionics that resemble those of fourth-generation fighters and a greater capacity for the use of guided weapons and sensors distributed across five hardpoints. Armoured, it has two internal 12.7mm (0.5in) machine guns, and also can operate with night-vision goggles and forward-looking infrared. It also has a digital cockpit, a head-up display, new ejection seats and encrypted communication systems. But the Super Tucano is another story.

Acknowledgements

The idea of writing this book arose in an informal conversation with Ary Guilherme Leber in August 2013, at Caçapava, São Paulo state. We were both motivated to tell this story to reflect the importance of the EMB-312 for Brazil and for the wider world.

Ary Guilherme Leber opened many doors so I could find the necessary information, people and data to help piece together this book. Colonel Aparecido Camazano wan as indispensable help and spared no effort to assist me. He not only opened files on the Tucano but also on the Brazilian Air Force itself and the history of national aviation.

In almost four years of dedicated research there were many moments when I was forced to be away from my wife, Nathaly Fonseca Nunes, and then from my little daughter, Amália. I had to leave them for days to dedicate myself to researching and writing this work. Even with these sacrifices, I was supported and encouraged to continue at all these times. Without you, Nathaly, I would not have got this far.

The research for a book of this scope is sometimes arduous and I could not fail to mention a central element in the development of this work. My friend Murilo Camargo

Martins, a distinguished illustrator, has worked extensively to recreate in coloured profiles most of the paint schemes and ordnance configurations used by 16 countries. Since August 2013, when it all began, he entered into this endeavor with sheer enthusiasm and passion for the Tucano. The constant exchange of emails and phone calls was sometimes the fuel that encouraged me to continue. Thank you very much, Murilo.

Below, in alphabetical order, I list all those who helped me in this project and without whom I would not have been able to finish this work. All of them, without exception, were enthusiastic and readily available to assist. Unfortunately, there are some names I cannot include and others to whom I cannot give due acknowledgement for their efforts. Suffice to say, their contributions were equally important.

1°/3° GAV, 3°/3° GAV, 1° Grupo de Defesa Aérea , Aparecido Camazano Alamino (Colonel, Brazil), Armée de l'Air, Frank van der Avoort (Netherlands), André Luiz Bandeira (Brazil), Johnson Barros (Sergeant, Brazil), Sargnon Bastien (France), Bruno Batista (Sergeant, Brazil), Flávio Flores da Cunha Bierrenbach (Brazil), Farzad Bishop, Força Aérea Brasileira, Derek Bower (United Kingdom), Michael Brazier (United Kingdom), Neil Bury (United Kingdom), Luiz Fernando Cabral (Colonel, Brazil), Mário Roberto Vaz Carneiro (Brazil), Willie Monteiro R. de Carvalho (Colonel, Brazil), Leandro Casella (Brazil), Erwan de Cherisey (France), Tom Cooper (Austria), Fuerza Aérea Colombiana, Corporación de la Industria Aeronáutica Colombiana (CIAC, Colombia), Gustavo Cury (Captain, Brazil), Marcelo Kanitz Damasceno (Brigadier, Brazil), Simone Dantas (Lieutenant, Brazil), Tom Dean (United Kingdom), José Ricardo Drozdz (Brazil), André Oliveira Duailibi (Brazil), Flavio Enrique Ulloa Echeverry (General, Colombia), Oscar Elvir (Honduras), Embraer, José Faria (Colonel, Brazil), FAV-Club (www.fav-club.com), Renzo Galuppo Fernandes (Brazil), Gabriel Fioni (Argentina), Roberto Follador (Colonel, Brazil), Antônio Luis Sapienza Fracchia (Paraguay), Leonardo Garcia Gaedke (Sergeant, Brazil), Joop de Groot (Netherlands), Márcio Inforzatto (Lieutenant, Brazil), Cristóvão M. da Silva Júnior (Admiral, Angola), Waldemar Prieto Júnior (Sub-officer, Brazil), Ali Reza Khodakarami (Iran), Joseph Kovács (Hungary), Maurício Lanza (Brazil), Ary Guilherme Leber (Brazil), Humberto Leite (Lieutenant, Brazil), Chris Lofting (United Kingdom), Ahmad Mahgoli (Iran), Paulo Marques (Sergeant, Brazil), Mário Sérgio Greskow Martinhão (Colonel, Brazil), Lauro Ney Menezes (Brigadier, Brazil, in memoriam), Mauro César Mezzacappa (Brazil), Luiz Alberto C. Munaretto (Colonel, Brazil), Rinaldo Nery (Colonel, Brazil), Ivan P. Nesbit (Venezuela), Luiz Fernando Nolf Ventura (Brazil), Rogério Olegário (Colonel, Brazil), Sérgio Barros de Oliveira (Colonel, Brazil), Carlos Filipe Operti (Brazil), Giuseppe Ricardo Passarini (Brazil), Fuerza Aérea del Peru, João José Pereira Pinto (Colonel, Brazil), Luciano Porto (Brazil), Lewis Mejia Prada (Peru), Marco Aurélio do Couto Ramos (Brazil), Paulo Rezende (Sergeant, Brazil), Santiago Rivas (Argentina), Andres Esteban Robles (Argentina), Lars Rohwer, Ricardo Magno Silva Rosa (Colonel, Brazil), Marc Rosenholtz (Belgium), Thiago Sabino (Brazil), Chris Sacco (France), Parque de Material Aeronáutico de Lagoa Santa (Brazil), Sérgio Santana (Brazil), Shahram Sharifi (Iran), Ozires Silva (Colonel, Brazil), Rogério Terlizzi (Brazil), Amaru Tinopa (Peru), Javier Franco Topper (Colombia), Celso Cardoso Vilarinho (Major, Brazil), Mário Vinagre (Brazil), Instituto de Pesquisas e Ensaios em Voo (Brazil) and Pit Weinert (Germany).

João Paulo Zeitoun Moralez, October 2017

Abbreviations

AdA	Armée de l'Air (French Air Force)
ADF	automatic direction finder
AEW&C	airborne early warning and control
AFA	Academia da Força Aérea (Brazilian Air Force Academy)
ALADA	Ala de Defesa Aérea (Brazilian Air Defence Wing)
AMB	Aviación Militar Bolivariana (formerly Venezuelan Air Force)
BAC	British Aircraft Corporation
BAVEX	Batalhão de Aviação do Exército (Brazilian Army Aviation Battalion)
BIS	Batalhão de Infantaria de Selva (Jungle Infantry Battalion)
CATRE	Centro de Aplicações Táticas e Recompletamento de Equipagens (Brazilian Tactical Applications and Equipment Recomplement Centre)
CTA	Centro Técnico Aeroespacial (Aerospace Technical Centre)
COMAT	Comando Aerotático (Tactical Air Command)
COIN	counter-insurgency
DCTA	Departamento de Ciência e Tecnologia Aeroespacial (Brazilian Department of Aerospace Science and Technology)
DME	distance measuring equipment
EAF	Egyptian Air Force
EB	Exército Brasileiro (Brazilian Army)
ELO	Esquadrilha de Ligação e Observação (Liaision and Observation Squadron)
FAA	Fuerza Aérea Argentina (Argentine Air Force)
FAB	Força Aérea Brasileira (Brazilian Air Force)
FAC	Fuerza Aérea Colombiana (Colombian Air Force)
FAH	Fuerza Aérea Hondureña (Honduran Air Force)
FAIM	Force Airienne Islamique de Mauritanie (Mauritanian Air Force)
FAP	Fuerza Aérea Paraguaya (Paraguayan Air Force)
FAPe	Fuerza Aérea del Peru (Peruvian Air Force)
FANA	Força Aérea Nacional de Angola (Angolan Air Force)
FARC	Forças Armadas Revolucionárias Colombianas (Revolutionary Armed Forces of Colombia)
FAV	Fuerza Aérea Venezoelana (Venezuelan Air Force, later AMB)
FIDA	Feria Internacional del Aire
GAVCA	Grupo de Aviação de Caça (Brazilian Fighter Squadron Group)
GAV	Grupo de Aviação (Aviation Group, Brazil)
GDA	Grupo de Defesa Aérea (Air Defence Group, Brazil)
CINDACTA	Centro Integrado de Defesa Aérea e Controle de Tráfego Aéreo (Brazilian Integrated Centre for Air Defence and Air Traffic Control)
HOTAS	hands on throttle and stick
HUD	head-up display
IrAF	Iraqi Air Force
IFR	instrument flight rules
ILS	instrument landing system
IRGC	Islamic Revolutionary Guard Corps

IPEV	Instituto de Pesquisas e Ensaios em Voo (In-Flight Research and Testing Institute)
ITA	Instituto Tecnológico de Aeronáutica (Technological Institute of Aeronautics)
KAF	Kenya Air Force
KAF	Kuwait Air Force
km	kilometres
km/h	kilometres per hour
MAP	Military Assistance Program (US)
MASL	metres above sea level
MB	Marinha do Brasil (Brazilian Navy)
MRO	maintenance, repair and overhaul
NATO	North Atlantic Treaty Organization
nm	nautical miles
NVG	night-vision goggles
OBOGS	on-board oxygen generating system
PEF	Pelotão Especial de Fronteira (Brazilian Army Special Border Platoon)
RAF	Royal Air Force (UK)
rpm	rotations per minute
SIVAM	Sistema de Vigilância da Amazônia (Amazon Surveillance System)
TCAS	traffic collision avoidance system
USD	United States Dollar
USAF	United States Air Force
USN	United States Navy
VOR	very high frequency omnidirectional range

EMBRAER'S PIONEER YEARS

In 1968, at the heart of the then Centro Técnico Aeroespacial (CTA, Aerospace Technical Centre), today the Department of Aerospace Science and Technology, the IPD-6504 prototype took shape inside hangar X-10 in São José dos Campos – a major city in São Paulo state. This project later became known as the Embraer C-95 Bandeirante – in its military version – and the EMB-110 Bandeirante in its passenger transport version.

Embraer did not yet exist, but a series of events came together and led to its creation the following year. Those events consisted mainly of political decisions, by officers of the Força Aérea Brasileira (FAB, Brazilian Air Force) and other visionaries, who wanted to equip Brazil with its own aeronautical industry at all costs. This was a moment of great excitement, and to understand it we need to consider several factors that occurred simultaneously.

Production of the Bandeirante aimed to serve the domestic regional aviation segment that lacked a modern, robust and efficient aircraft that would operate on the low-demand routes, replacing the ageing Douglas C-47/DC-3.

Aviation was developing at a pace. The introduction of the turbine engine saw airlines increasingly adopt jet aircraft, which offered more comfortable flying, were faster and had better overall performance. However, they were also more expensive to purchase, operate, and required long paved runways and complex ground infrastructure to support each take-off and landing. Moreover, to be economically viable, since the aircraft were designed for routes of greater density and had many seats, they required a greater occupation rate on each flight.

These two crucial points led to smaller towns and cities no longer being served by air, since they lacked a suitable volume of passengers and the infrastructure necessary to make these operations feasible. At such locations in Brazil, landing strips were generally short stretches of dirt or grass, there was no aviation-grade kerosene available (only gasoline for piston engines) and many key support items – including stairs for the passengers – were missing.

A few companies continued flying to these cities, but it was a precarious existence, since they were operating aircraft that did not meet the requirements adequately or efficiently.

The Bandeirante was the complete antithesis of the jet airliner and for that reason proved to be a revolutionary development for the regional aviation sector. It had a 15-passenger capacity (later versions could carry up to 19), was strong, fast, could operate from unprepared runways and had low acquisition and operation costs. When it entered the market in the mid-1970s, the Bandeirante gained huge acceptance and filled a long-standing gap across the globe.

Development of the twin-turboprop C-95 Bandeirante was begun by the Brazilian Air Force and was transferred to Embraer in 1970, when the prototype was used in flight tests. The aircraft became the backbone of the FAB's light tactical transport fleet and helped commercial aviation to develop routes to small and far-flung cities.
(Sgt Batista, FAB)

However, the Brazilian Air Force did not have the resources to finance the programme, so it was necessary to seek the support of other interested parties for its production. At the same time, the FAB was going through a difficult period as regards its fighter aviation arm. In fact, in the second half of the 1960s, it virtually ceased to exist, thanks to the early deactivation of 60 Gloster Meteor F.Mk 8 fighters and 10 Meteor T.Mk 7 two-seat trainers. The factory-fresh aircraft had begun to be delivered from the United Kingdom in 1952, to equip the 1º Grupo de Aviação de Caça (GAVCA, 1st Brazilian Fighter Aviation Group), at Santa Cruz Air Base, in Rio de Janeiro, and the 1º/14º Grupo de Aviação 'Esquadrão Pampa' (GAV, 1st/14th Aviation Group 'Pampa Squadron') at Canoas Air Base in Rio Grande do Sul. The Meteor was a modern and advanced platform for its time – it was also the first jet operated in Brazil.

Nevertheless, inspections in 1965 indicated serious structural failings that made operations unsafe. The 'Pampa Squadron' was the first to retire the jet, and by 1968 its deactivation was already at a very advanced stage. As a result, the air defence mission was assigned to the Lockheed T-33, an aircraft completely unsuited to the job, since it had no heavy weapons or adequate sensors for that kind of mission. The study, analysis and purchase of a new aircraft became the FAB's priority, and at the same time the air force would initiate the implementation of a complex radar network for airspace monitoring and surveillance.

There were other problems too. In an informal meeting with Colonel Lauro Ney Menezes, one of the most brilliant and visionary minds ever to have served the FAB, Lieutenant-Colonel Ozires Silva, who later became Embraer's president, explained that the training aircraft of the Academia da Força Aérea (AFA, Air Force Academy) were entering an irreversible process of obsolescence. In addition, the North American T-6, which also performed ground attack, armed reconnaissance and tactical support missions, was at the end of its lifespan.

A short time later, Menezes, Ozires Silva and Major Ivan Moacyr da Frota were tasked by Brigadier Márcio de Souza e Mello, then Minister of Aeronautics, with a mission to operationally evaluate the BAC Lightning (United Kingdom), Dassault Mirage IIIE (France), Lockheed F-104 Starfighter (built under licence by FIAT Aviation in Turin, Italy) and the Saab Draken (Sweden). Aware of the FAB's deficiency in tacti-

In production since 1970, the Ipanema was formulated by the Brazilian Air Force in response to a Ministry of Agriculture requirement for an agricultural aircraft. The programme was later transferred to Embraer. (Embraer)

cal aircraft, they decided to ask the companies to evaluate the possibility of granting Brazil a licence for the production of their advanced training jets, to be used for operational conversion to front-line fighters and also in air support missions. The UK offered the possibility of building the BAC 167 Strikemaster under licence. Sweden offered the Saab 105, France the Fouga Magister and Italy the Aermacchi MB-326G.

Back in Brazil, the plan was presented and approved by Minister Márcio de Souza e Mello, who, in addition to the fighter and training aircraft, provided for the establishment of a complex radar network for the surveillance and control of Brazilian airspace, thereby increasing sovereignty and flight safety. In the future, this radar network became known as the Centro Integrado de Defesa Aérea e Controle de Tráfego Aéreo (CINDACTA, Integrated Centre for Air Defence and Air Traffic Control). Of course, together these programmes were extremely expensive to fund and implement. Nevertheless, the equation was almost solved.

On 26 June 1969, before the creation of Embraer, a meeting was held with Ministers Delfim Netto (Finance), Hélio Beltrão (Planning) and Macedo Soares (Industry and Commerce), at which the programme for the acquisition of fighters and construction of the Bandeirante was presented, together with licence production of the advanced jet trainer. At the time there was already a political inclination for Brazil to develop

The EMB-400 Urupema was a high-performance glider designed by Guido Pessotti. Ten were built using Embraer funds after contracts had been signed for the Bandeirante and licence production of the EMB-326GB Xavante. (Valdemar Júnior)

The Embraer EMB-326GB Xavante was widely used in the FAB's COIN and reconnaissance squadrons and also to qualify and convert pilots to front-line jet fighter squadrons. (Embraer)

its own aeronautical industry, which was deemed of strategic importance and fundamental for the country to develop critical technologies and highly specialised human resources. If private initiatives did not invest in the Bandeirante project, it would be necessary to resort to government funding.

The benefits of that meeting were soon reaped with the signature of Decree Law No. 770, on 19 August 1969, by Marshal Artur da Costa e Silva, then President of Brazil. Embraer – Empresa Brasileira de Aeronáutica S.A. – was formed as a government-controlled company, with the Federal Government holding 51 per cent of the shares and the remaining 49 per cent resulting from shares of individuals and legal entities. The company only started operations on 2 January 1970, when several actions were carried out, before being consolidated in the signature of two major contracts in May 1970.

The first contract covered the manufacture of 80 twin-engine C-95 Bandeirante turboprops for the FAB. These would be deployed among the various aerial transport units, flying missions in any part of the national territory. Among the aircraft developed by Embraer, the Bandeirante still holds the title for the largest number of versions, including aerial photogrammetry, photo-reconnaissance, maritime patrol and surveillance, aerial laboratory and radar calibration, along with the utility transport versions themselves. The second contract provided for the production, under licence, of 112 MB-326GB advanced training and light attack jets. Designated by the FAB as the AT-26 Xavante, the jet would be responsible for converting new military pilots to fly the supersonic fighters that soon would be acquired through an international tender opened by the FAB. They would also be used for ground-attack missions, close air support and escort.

Also in May a contract was signed for the acquisition of 12 single-seat Mirage IIIEs and four two-seat Mirage IIIDs. The classic French delta had already made history by being a leading protagonist in the Six-Day War of June 1967, a conflict in which Israel effectively annihilated four Arab air forces, including the powerful Egyptian and Syrian formations, in a matter of a few hours. Starting in 1972 these 16 aircraft equipped the 1ª Ala de Defesa Aérea (1ª ALADA, 1st Air Defence Wing), in 1979 transformed into the 1° Grupo de Defesa Aérea (1° GDA, 1st Air Defence Group), at Anápolis Air Base, in Goiás.

The EMB-121 Xingu was the first pressurised-cabin project developed by Embraer. The twin-engine executive aircraft had nine seats and was intended for short-haul services.
(Cb Silva Lopes, FAB)

At the time, the two contracts – C-95 and MB-326 – were worth USD 1.2 billion, which also provided the company with the resources necessary to manufacture 10 EMB-400 Urupema high-performance gliders, a project of the aeronautic engineer Guido Pessotti. A graduate of the Instituto Tecnológico de Aeronáutica (ITA, Technological Institute of Aeronautics), while working there as a teacher, Pessotti was incorporated into the Bandeirante programme at a time when it was still at the CTA as the IPD-6504 prototype. He began to directly assist Max Holste, a French aeronautical constructor who had arrived from Europe to manage the twin-engine turboprop programme at the CTA. The qualities and innovations seen in Embraer's aircraft were almost unique in comparison to similar models from other international manufacturers, mainly due to the brilliant mind of Pessotti, who always strived to meet the demands of the operators and to offer the market, at the right time, projects that would meet the demands of aviation operators around the world.

Also reassigned to Embraer was the EMB-200 Ipanema, a project developed by the CTA as the IPD-6909 to meet the requirements and specifications of the Ministry of Agriculture. This was an agricultural aircraft capable of operating for long periods without the need for complex infrastructure. With the beginning of operations at Embraer, the project was transferred to the company's responsibility, and the prototype's inaugural flight was accomplished on 31 July 1970.

Certification was achieved in December 1971 and by January the following year several aircraft had already been delivered. The single-engine Ipanema continues to be manufactured by Embraer to this day, having received dozens of modifications and improvements to meet the most varied requirements of the market – among them, operation with automotive ethanol instead of aviation gasoline, which makes it less polluting and cheaper to operate.

On 7 September 1971, Embraer delivered the first of two AT-26 Xavantes to the FAB, while on 1 August 1972 the first Bandeirante series aircraft was rolled out for its maiden flight. On 20 December of that year the Bandeirante was certified and on 9 February the following year the FAB received its first two aircraft. Meanwhile, Transbrasil, a Brazilian company, was the first commercial airline in the world to receive Embraer's twin-engine turboprop.

The newly created Brazilian company was fully aware that it would be impossible to survive only on domestic orders – which came mainly from the government to equip the FAB. With this in mind, Embraer decided to pursue two initiatives. First, it developed versions of the EMB-110 Bandeirante oriented towards the external market, with the stretched EMB-110P1 and P2 versions. These were a great success on the international market, selling to operators in five continents. This helped to cement the Embraer brand in the international market, which, of course, contributed greatly to future sales of the Tucano. The second initiative aimed to secure the long-term future of the company via diversification. In this way, it sought to establish partnerships, with various agreements aimed at the international market.

In August 1974, Embraer signed an important agreement with Piper, the established US manufacturer, for the construction under licence in Brazil of some of its models – including the twin-engined EMB-820 Navajo and EMB-810 Seneca and the single-engined EMB-710 Carioca, EMB-711 Corisco, EMB-712 Tupi, EMB-720 Minuano and EMB-721 Sertanejo – to fulfill the needs of the light aviation sector. The Brazilian manufacturer also benefited from an offset agreement when Brazil purchased a batch of 36 Northrop F-5E Tiger II fighters and six F-5B Freedom Fighter trainers. Embraer was commissioned to produce 150 vertical empennages and wing pylons for the F-5E, after training its employees for this work.

The company continued to pursue other market developments, and on 20 October 1976 the EMB-121 Xingu, Embraer's first pressurised aircraft, took off on its inaugural flight, establishing yet another important milestone in the company's industrial chain in terms of technological development. It resulted in just over 100 airframes produced.

Embraer has always been a proactive company, looking for products and solutions that could meet the future needs of the market. The Bandeirante is evidence of this. Developed at the CTA, its design was based on the concept of a passenger aircraft that could handle regional transportation demands, complementing airlines' existing operations by connecting small cities to large regional hubs, where passengers could find a large number of connections flown by jets to dozens of destinations.

Embraer's turboprop airliner took eight years to develop, certify and then enter service with an airline. The interval between conception and regular operation may seem long and even suggest that the company missed the best moment to launch the product on the market. However, Embraer's farsightedness and provisions ensured the Bandeirante was made available to airlines just as they were searching for an aircraft with those exact characteristics.

Like the Bandeirante and other products of the company, the EMB-312 Tucano was developed with a view to meeting the FAB's needs, in this instance to replace the Cessna T-37C jets used by the Air Force Academy. As such, this was another opportunity identified by Embraer directors, this time in the late 1970s. An extremely short timeline was selected to develop the turboprop, which eventually equipped 16 air forces and saw service worldwide due to its versatility and robustness. Even so, with enthusiasm, passion and competence, the Embraer team spared no effort to ensure the project not only became a reality, but also a breakthrough in the field of military flying training.

Altogether, in addition to Brazil, 464 EMB-312s have been operated by Angola, Argentina, Colombia, Egypt, France, Honduras, Iran, Iraq, Mauritania, Paraguay, Peru and Venezuela. In the version produced under licence by Short Brothers in Northern Ireland, the Tucano has also served in the United Kingdom, Kenya and Kuwait, with a total of 158 units operated by these three countries.

During a period of only 10 years, the EMB-312 Tucano was fielded by almost all of its eventual operators. Here, aircraft are seen prior to delivery to (front to back) Venezuela, Peru, Brazil and Argentina. (Embraer)

ROADMAP TO THE TUCANO

The FAB training model

With the creation of the Ministry of Aeronautics on 20 January 1941 the Military Aviation (of the Brazilian Army) and Naval Aviation (of the Brazilian Navy) were disbanded, and most of their personnel and all their aircraft were transferred to the then newly created Força Aérea Brasileira. In this way, two former educational institutions, the Escola de Aviação Militar (Military Aviation School, which became the Military Aeronautic School in 1938 and the Army Aeronautic School in 1940) and the School of Naval Aviation, were also closed down, and the School of Aeronautics was created through Decree Law No. 3.142, dated 25 March 1941.

The new educational institution occupied the premises of the former Army Air Force School at Campo dos Afonsos, in Rio de Janeiro.

The site, however, was very precarious for aerial instruction due to the mountainous geography around the airfield, inadequate meteorology for flights and due to the increase in the already massive air traffic around Rio de Janeiro – at that time, federal capital of Brazil.

On 23 January 1942 a group of flying officers was assembled to find a location for the construction of a new headquarters for the School of Aeronautics. Candidate cities were required to have good infrastructure, be close to large urban centres, have good communication routes and ease of access, be far from foreign borders and have a flat topography, among other features.

The São Paulo and Minas Gerais Plateau was considered more suitable, especially in the region between Campinas and Ribeirão Preto, in São Paulo state; and between Belo Horizonte and Pirapora, in Minas Gerais state. Viracopos (Campinas), Campo Redondo (Rio Claro), Campos Altos (Pirassununga) and Lagoa do Campo (Ribeirão Preto) were all on the shortlist. Finally, on 18 December 1942, Pirassununga was chosen to house the new school.

The move from Rio de Janeiro to São Paulo began in the early 1960s. In 1964 air operations commenced in Pirassununga. At this time, the students, still in Campo dos Afonsos, had flown the already obsolete Fokker S 11 and S-12 (designated T-21 and T-22 in the FAB, respectively) during primary training, passing through advanced instruction in São Paulo in the venerable North American T-6D/G.

On 10 July 1969 the School of Aeronautics was reassigned to the Air Force Academy and on 23 October 1971 the final move was made from Rio de Janeiro to Pirassununga.

In the mid-1960s, the Ministry of Aeronautics launched a competition for replacement of the T-6, which already showed signs of obsolescence in the role of advanced

Still in use for basic training at the Air Force Academy, the Neiva T-25 Universal emerged from a project for a basic and advanced trainer with weapons capability for the Brazilian Air Force. It was designed by Joseph Kovács, responsible for many COIN aircraft projects. The first studies for the future EMB-312 used the fuselage and tail of the T-25.
(João Paulo Moralez)

trainer. Two national companies entered the competition with different projects, Neiva with the N-621 Universal and Avibras with the A-90 Alvorada.

Neiva's N-621 was equipped with a Hartzell constant-speed two-blade metal propeller and a 300hp Lycoming IO-540-K1D5 engine. It had a maximum speed of 300km/h (161kts), range of 1,500km (809nm) and operational ceiling of 6,100m (20,000ft). Avibras, which was created in part to meet this requirement of the Ministry of Aeronautics, presented the A-90 Alvorada. This had capacity to carry up to four people, a cruising speed of around 220km/h (118kts), a low wing and endurance of four hours. The N-621, which would enter service as the T-25 Universal, was far superior to the A-90 Alvorada and was chosen by the FAB.

Production of the Universal began in May 1965 and the type flew for the first time on 29 April 1966, in São José dos Campos. Receiving the civil registration PP-ZTW, the prototype was flown by test pilot Brasílio Freire Neto. In December 1967 the FAB chose to purchase 150 examples of the T-25, the low-wing single-engine aircraft designed by one of the world's greatest aeronautical engineers – the Hungarian Joseph Kovács. Although the order was reduced to 140 aircraft, this was the largest purchase in the history of Brazilian aeronautical industry to that date. The first T-25s were received in 1971 and were assigned to various units of the FAB to perform administrative, utility, reconnaissance, attack and liaison missions.

In 1966, prior to the arrival of an indigenous T-6 replacement for advanced instruction, and once T-6 fleet availability became so low that it affected flying training, the FAB decided to buy a batch of 40 Cessna T-37C advanced jet trainers. In 1967 these were delivered direct to Pirassununga, for advanced instruction. In the same year, the T-6 was finally replaced in the advanced pilot training role. But it was not just the advanced trainer that was obsolete. There was also a need to replace the old Fokker

T-21 and T-22 whose lifespan had come to an end and which were unsuitable for the evolving demands of flying training.

Once more seeking to encourage the national aviation industry, the FAB acquired 76 Aerotec A-122A Uirapurú basic training aircraft in October 1967, these being powered by a 160hp Lycoming O-320-B2B piston engine. Delivery of the aircraft began in 1969.

In the late 1960s the Ministry of Aeronautics implemented a new doctrine in the training of its new cadets – the Curso de Formação de Pilotos Militares (CFPM, Military Pilots Training Course). Until then, cadets at the AFA had completed the Aircraft Training Course over a period of four years, including primary, basic and advanced instruction. It was not uncommon, especially after the adoption of the T-37C, for a significant percentage of the cadets to fail to meet the requirements of the course or simply not get used to the peculiarities of military flight. As well as involving lost time for the cadet himself, the FAB ended up making significant investments in students who did not finish the course.

Beginning in 1970, those who joined the AFA to become military pilots were first sent to Natal, in Rio Grande do Norte, where they would receive primary and basic instruction in the T-23 Uirapurú (which had only recently been received, replacing the T-21 and T-22). They then underwent advanced training in the T-37C, as well as academic subjects related to aviation including meteorology, aerodynamics, aerial navigation, intelligence and air traffic, among others. At the end of this phase, the cadet went to Pirassununga to receive the remainder of their military training for a period of three years, while continuing to fly the T-37C only to maintain proficiency.

With this new demand the FAB acquired a second batch of 25 T-37Cs that were exclusively destined for the CFPM in Natal. In 1973 the T-25 Universal replaced the T-37C in Natal, in the advanced training role, on which occasion the jets were concentrated at the AFA, and the following year CFPM was terminated. Despite having only a brief existence, the CFPM was the most important development implemented by the FAB in the field of military flight instruction. In 1974 flight instruction was centred at Pirassununga with the T-23 and T-37C trainers.

The T-25 continued operating in Natal at the former Centro de Aplicações Táticas e Recompletamento de Equipagens (CATRE, Tactical Applications and Equipment Recomplement Centre), schooling second-class reserve officers, before being replaced in 1978 by the C-95 Bandeirante. This replacement was necessary so that the T-25 could assume advanced training duties at the AFA, since the T-37C was in the process of early deactivation due to the termination of a military agreement with the US, which brought to an end the supply of spare parts.

Nuclear programme and birth of the Tucano

During the 1930s and 1940s, in the context of the study and exploration of nuclear energy, Brazil was nothing more than an exporter of raw materials for trials and projects conducted in the US. At the beginning of the 1950s, motivated by the national nuclear policy written in 1947 by Navy Officer Alvaro Alberto da Mota e Silva, the Brazilian government created the Conselho Nacional de Pesquisas (CNPq, National Research Council), an institution dedicated to the study and development of knowledge in the most diverse areas, but mainly focusing on nuclear energy.

In this field, the then West Germany was a strong partner for Brazil. In the mid-1950s the government received three research centrifuges from Germany, which entered service in 1970. At that particular time the subject was already dealt with on a strategic level, although few achievements had been achieved in practical terms. With the election of Juscelino Kubitschek to presidency the country adopted a more nationalistic position regarding nuclear development. The Comissão Nacional de Energia Nuclear (National Nuclear Energy Commission) was created and all contracts that foresaw the export of nuclear minerals (such as monazite, thorium and uranium) were cancelled.

The US contributed by providing Brazil with a nuclear research reactor, the IEA-R1, through the 'Atoms for Peace' programme in 1957. This was installed in the Institute of Atomic Energy (now known as the Instituto de Pesquisa de Energia Nuclear, IPEN, Nuclear Energy Research Institute) in São Paulo. It was the first of its kind in the southern hemisphere. In 1960 the TRIGA Mk 1 reactor was installed in the Radioactive Research Institute of the Federal University of Minas Gerais. Five years later, at the Institute of Nuclear Engineering, Rio de Janeiro, the first nuclear reactor built in Brazil – the Argonauta – became operational.

At that particular moment, coming after the 1964 Revolution that saw the introduction of a military government, the use of nuclear energy was considered strategic and it was considered essential for Brazil to be responsible for the entire process of production and handling of nuclear fuel. The technology would have civilian purposes, meaning the production of power. But as a consequence of its mastery of nuclear technology, Brazil could also develop, if it was in the country's interest, military technologies.

It is worth remembering that the Brazilian government opposed the Treaty on the Non-Proliferation of Nuclear Weapons and the limitation of peaceful nuclear explosions, as it believed that both measures limited sovereignty and national expansion plans in this area.

In the early 1970s, while Brazil signed a nuclear cooperation agreement with West Germany, details were concluded for the supply of a nuclear powerplant from Westinghouse, and supplies necessary for its operation, in coordination with the United States Atomic Energy Commission. In 1973 this political motivation was further strengthened by the oil crisis that took the world's powers of the day by surprise. In fact, by 1990, Brazil hoped to have 12 nuclear powerplants dedicated to generation of electric power, as foreseen in the so-called Plan 90.

The US, however, began limiting the export of nuclear material after India tested its own weapon in 1974. With the entry of Democratic President Jimmy Carter to the White House on 20 January 1977, Brazilian plans in this industry began to be threatened by the new policy adopted by the US.

The new president's stated objective and commitment was to reduce the existing arsenal of nuclear weapons and discourage other countries from arming themselves with such weapons.

In the specific case of Latin America and the Caribbean, the Treaty of Tlatelolco – as the Treaty for the Prohibition of Nuclear Weapons in Latin America and the Caribbean became known – came into force in 1968, a document which almost all nations of the region signed and joined. Despite this, both Brazil and Argentina were in a veiled nuclear race, with the two countries being supported by West Germany and Switzerland, respectively.

The Cessna T-37C had a short life with the Brazilian Air Force due to a lack of spares after Cessna stopped providing support for the Brazilian fleet. (Archive João Paulo Moralez)

Immediately, the US government initiated political measures and applied pressure to dissuade Brazil from its objective regarding the use of nuclear resources for military purposes. Official visits by Jimmy Carter to Germany and Switzerland at the same also addressed the support provided by those countries to Brazil and Argentina.

On 1 March 1977, US Deputy Secretary of State Warren Christopher was in Brasilia with proposals for Brazil to alter its policy regarding the use of nuclear power. Washington's efforts were in vain. The Brazilian government alleged that its use of nuclear technology was not intended for war, but only for peaceful purposes. Moreover, Brazil did not consider the pressure from the Carter administration to be coherent, since the US and other nations already in possession of this technology conducted their research and testing without any opposition.

Finally, on 11 March 1977, General Ernesto Geisel, then President of Brazil, unilaterally withdrew from the Military Assistance Program (MAP), a military agreement with the US signed on 15 March 1952, which had come into force on 30 April 1953. That stance was a direct response to the Americans and even the military was caught by surprise. Two reasons led to this decision on the part of Geisel. The first was linked to the nuclear issue, which in the eyes of the government ultimately challenged national sovereignty. The second, undisclosed, reason was directly related to US pressure on Brazilian political persecution and human rights abuse – another banner of the Carter Administration, which secured him the Nobel Peace Prize in 2002.

Under the MAP agreement, Brazil benefited from training, personnel exchanges and the purchase of relatively modern war materiel for the maintenance of national defence and sovereignty. In return, Brazil provided strategic raw materials for the US.

Within the FAB, several types of aircraft still in active service had been obtained through MAP channels. Among them was the T-37C. And it was at this moment that the first ideas emerged that culminated in the Embraer EMB-312 Tucano. As a way of retaliation, the US, via Cessna, sent a letter to the Aeronautics Command that would have repercussions for a branch of the military that had enjoyed high priority and official interest – the training of military pilots at the AFA. After all, the high quality of instruction provided to cadets at the academy was critical to the fielding of skilled and well-qualified flying officers. Ozires Silva, then president of Embraer, explained:

'There was indeed a restraint, perhaps because of US pragmatism towards us. I remember that we wanted to buy an IBM supercomputer to improve our calculation base, but the sale was vetoed.

'Around 1977, I was walking down a corridor of the Aeronautical General Staff, when Colonel Lélio Viana Lobo, who was my classmate at the School of Aeronautics in Campo dos Afonsos, Rio de Janeiro, exited a room and showed me a letter sent by Cessna. Its content said that for programming and production reasons the company would not be able to supply spare parts and technical support for the T-37C fleet beyond the next two years.'

The content of the letter did not disclose the true motive, which was the disruption of MAP. The aircraft were only 10 years old when the letter arrived, still very new despite being used for the training of cadets – an activity that naturally leads to greater wear and tear on the airframes. The situation could only be circumvented if the FAB bought large batches of spare parts, which was impossible due to budgetary restrictions. Without spare parts and logistical support, in a short time the T-37C fleet's availability would reach very low levels and it would begin to be scrapped – completely compromising the AFA's advanced training syllabus. The FAB now had to deal with a serious, complex problem and one that demanded an immediate solution.

In fact, in 1978, when the assistance provided by Cessna came to an end, the T-37C was retired from air training after only 12 years in service with the FAB. The maintenance issue proved to be extremely critical and availability had been declining, even with the FAB ground crews' commitment to keep more aircraft in flying condition. A few remained in use on utility flights until 1980, when they finally stopped flying.

Of the 65 T-37Cs purchased by Brazil, 42 were sold to South Korea, which operated them until the mid-2000s. Worldwide, countries including Colombia, El Salvador, Guatemala, Peru and Uruguay maintain operational A/T-37 fleets to this day. Until 2009, the aircraft was also part of the US Air Force training structure.

Ozires Silva continued:

'Because I was always 'on alert', I said, 'Lélio, you're going to need a training plane and we're going to do it!' *He told me to go to hell and said we did not know how to make training airplanes. He said a lot of things that I honestly do not even remember, because I was not listening to anything else. I was far away, thinking of the training plane. I will never forget that I went to Brasilia airport and returned on VASP Flight 291 that took off at 18.30 to São Paulo. The next day, I went early to Embraer's Engineering Department, met the section chiefs, and said,* 'We're going to make a training plane for FAB'. *The answer was:* 'Woo-hoo!''

Concept for a new trainer

Embraer's directors worked proactively and were always attentive to new and potential project opportunities to be developed and built within the company. This required constant contact with civilian, military and airline operators, private owners and air taxis, as well as analysis of new aviation niches and trends. So it was with the Bandeirante. The Tucano would be the next example of this.

'People in Brazil think that things happen automatically and I always like to emphasise the importance of people as decision makers and perceptions that change things as they are. And that's what happened at Embraer. We were always motivated, we always had a permanent enthusiasm for what we did, calling it a passion for our work. We were plugged in all the time. Most of Embraer's decisions came from the fact that we were prepared, so when the opportunity passed by we jumped on it', Ozires Silva recalled.

But there were fewer pros than cons.

To begin with, a trainer was not part of Embraer's plans (or those of the FAB, since the T-37Cs were essentially new). The team was too lean and overworked to concentrate on the various programmes that were already in progress. At that moment Embraer already had products that served general aviation, executive, sport, military, regional and even agricultural sectors. The company was growing and maturing, overcoming barriers, prejudices (including within the FAB) and strengthening. It constantly received new equipment and trained its employees in different disciplines.

The new aircraft would have to be developed from scratch and at the same time meet the strictest requirements of the FAB. The service wanted to equip its academy with the best and most modern aircraft in terms of advanced training. It would not be a simple or easy task, and above all it would not be quick. There was an urgent requirement, Embraer had not yet developed any training aircraft by itself, and the race against time would be relentless.

Perhaps the most fundamental challenge would be to convince the FAB's High Command that Embraer was capable of designing and producing a training aircraft in such a short timeframe and meeting the specifications of the Ministry of Aeronautics, which in turn had committed much of its budget to Embraer itself, on the Bandeirante, Xavante and Xingu, for example.

But although Embraer itself did not have experience of developing a training platform, it did have in its ranks a personality that was recognised as one of the most advanced minds in national aviation – Joseph Kovács.

The Hungarian settled in Brazil had an impressive aeronautical development curriculum, passing through such places as the Instituto de Pesquisas Tecnológicas (IPT, Aeronautics Division of the Technological Research Institute); the Convertiplano project at the CTA, as part of the team of Professor Henrich Focke (co-founder of Focke-Wulf); designing the collective pitch control of the first Brazilian helicopter, the Beija-Flor (Hummingbird) also at CTA; and at Neiva. At the latter, Kovács had made numerous contributions to national industry through the N-592 Regente and the T-25 Universal (N-621) basic and advanced trainer. In 1973 Kovács was already working at Embraer and at last, in 1977, a dream that had existed for more than a decade could finally start to come true.

Based on the design of the T-25, the engineer wanted to develop a larger variant of the aircraft that could be used as an attack platform. One of the ideas was the Bi-Universal, which would use a large portion of the T-25 airframe, with tandem seats, wingtip tanks and a powerplant comprising two 300hp Lycoming IO-540 piston engines. It had six hardpoints for weapons (four under the wings and two below the fuselage) and fixed guns in the nose. The project did not proceed.

Also on Kovács' drawing board was the Universal II that would have the same wings and fuselage as the T-25 but with a 400hp Lycoming IO-720 engine and three-blade propeller. This design was also known as the N-621A.

Taking advantage of the wings, rudder and engine of the Universal II – but changing the cockpit configuration from side-by-side to tandem seats and including wingtip tanks – produced the N-721A, known as the Carajá. The landing gear was of the retractable tricycle type and the aircraft had a fuselage of all-metal construction. There was also the N-721B that was identical to the N-721A, except for its 550hp Pratt & Whitney PT6A-20 turboprop engine.

Ordnance load of the N-721 would comprise a maximum of 550kg (1,212lb) of stores, allowing the use of 7.62mm (0.3in) machine gun pods, 113kg (250lb) bombs, SBAT 70 (70mm, 2.75in) or SBAT 37 (37mm, 1.45in, training) rockets on four wing pylons. Top speed was 340km/h (183kts).

While at Embraer Kovács had decided to return again to a counter-insurgency (COIN) aircraft, examining the possibility of taking advantage of the EMB-110 Bandeirante propulsion set, wings based on those of the Navajo, and landing gear, control surfaces, parts of the central section and outer panels of the wing and rear fuselage from the T-25. Kovács considered that further modifications would be necessary but, even so, the company would gain development time and money.

The cockpit would have an arrangement similar to that of the AT-26 Xavante, with tandem ejection seats and a single-piece canopy opening to the right.

The project was presented in September 1976 in the report entitled '312-DV-00 Tactical Support Aircraft/COIN-X/TX Military Trainer'.

The factor that prevented any of these projects from advancing further was the lack of interest from the Ministry of Aeronautics. At the time there was simply no operational need, since the flight lines of FAB combat squadrons were being flooded with large numbers of AT-26 Xavantes. These, along with the T-25s, conducted special COIN missions in the reconnaissance and mixed attack squadrons.

However, despite the new urgency and the short period of time available for the creation of the trainer, developing a new platform would not be a problem for Kovács, who already had in his portfolio several projects dedicated precisely to this function (advanced trainer, and light attack).

For Embraer, which had already established its technical and industrial capabilities, a professional like Kovács was the missing link required for the programme to be developed as quickly as possible, taking into account the tight schedule demanded by the Ministry of Aeronautics.

The first studies

Until this point, the primary advanced training aircraft produced by different Western and Eastern manufacturers had two predominant configurations, employing either side-by-side or tandem twin seats. These designs included the Aero L-29, BAC Jet Provost and Strikemaster, Canadair CT-114 Tutor, CASA C-101 Aviojet, Cessna T-37, Dassault/Dornier Alpha Jet, Fouga Magister and Hawker Siddeley (later BAe) Hawk, which had been selected according to the operational needs or doctrine of each respective country.

One point on which these countries agreed was that the use of jet aircraft was indispensable for the task of advanced training. They believed that turbine powerplants ensured the training of their pilots would be more complete and effective. The exception to this rule was found in aircraft such as the Pilatus PC-7, for example, a Swiss-made tandem two-seat turboprop that could also be used for ground-attack missions.

Several aspects in this discussion supported the use of jet aircraft, among them overall performance, including the ability to perform higher-performance aerobatic manoeuvres and fly at higher speed – resulting in the need for decision making and actions taken in fractions of seconds. The cockpits of most of these aircraft had ejection seats and avionics very similar to those seen in third-generation fighters. With these characteristics, and based on the doctrines of the time, pilots would train in an environment as close as possible to the supersonic fighter aircraft that awaited them.

On the other hand, there were some disadvantages to the jets. The costs of acquisition and especially of maintenance were very high over the years. What made this reality even worse was the first oil price surge in the 1970s, when the cost jumped from a mere USD 3.00 per barrel to almost USD 12.00 in 1973 and around USD 40.00 in 1979. The crisis had been deliberately triggered by member countries of the Organization of Arab Petroleum Exporting Countries (OPEC) in response to the US involvement in the 1973 October War.

Ozires Silva continued:

'What are the key features of this training airplane? In the aeronautical industry we have to make the specifications of the product, of what we are going to manufacture, whether it is high or low wing, if it is aerobatic, the acceleration, ultimately, all those attributes that are inside the airplane have to be expressed together with the solutions we are proposing. I called Guido and said that we had to sketch a few things about this new airplane and later talk with all the team, about differentials that could make it competitive and win a prominent position in the world market. That was because we were already convinced that Brazil alone would never support a company the size of Embraer, which continues to export most of its production to the present day. The international environment was necessary. So our airplane had to have differentials.'

Embraer, aware of the advantages and disadvantages of the different types of advanced trainer, opted to develop a more conventional platform. In its first configuration this was powered by a supercharged 380hp Lycoming TIO-541-E1C4 piston engine driving a three-blade variable-pitch propeller.

The project was presented in May 1977 under the designation EMB-301 in study EP-301/105 for an advanced general training aircraft that could, in addition to being operated by the FAB, be exported to developing nations, reflecting its low cost of acquisition and operation and its high degree of operational flexibility.

Advanced training, the primary mission of the aircraft, would include formation flying, aerobatics, instrument flight rules (IFR) flying and aerial gunnery. Secondarily, the aircraft was intended for operation by the reconnaissance and attack squadrons, as well as liaison and observation squadrons.

For this, the aircraft required the capability to stay airborne for long periods, a high degree of operational availability, simplified maintenance, low dependence on ground support equipment, operation from short and unpaved runways, and good visibility thanks to its single-piece windshield and canopy. The tandem two-seat configuration would allow for the installation of ejection seats (optional), and the EMB-301 featured dual controls and instrument panel architecture similar to the fighters of the time.

The wings would accommodate two 215-litre (56.7-gal) main fuel tanks, plus two leading-edge tanks holding 210 litres (55.5 gal) each, for a total of 850 litres (224.5 gal). Under the wings, four hardpoints would allow carriage of a 1,360kg (3,000lb) load of ordnance such as free-fall, fragmentation or incendiary bombs, drop tanks, 7.62mm (0.3in) machine gun pods, 37mm (1.45in), 57mm (2.24in), 68mm (2.67in), 70mm (2.75in) or 127mm (5in) rocket launchers and reconnaissance pods.

The EMB-301 was compared to 16 aircraft (four piston-engined, five turboprops and seven jets): these included the MB-326KB, Strikemaster, Beechcraft T-34 Mentor, T-37, EMB-326 Xavante, FMA IA-58 Pucará, Magister, Neiva T-25 and Rockwell OV-10 Bronco, among others. However, it was clear to all at Embraer that the EMB-301, being a more refined variant of the N-721A Carajá, would not seriously attract the attention of the Ministry of Aeronautics. It was necessary to present a project that would provide more innovations for the advanced training role.

The next response to Embraer's own criticism was the EMB-311 and, if time and money were Embraer's priorities, this configuration was the one that best utilised parts and components of other projects to be able to fulfill the short development time required by the FAB.

The EMB-311 was featured in project EP-311/104 (second configuration), this being an attack aircraft to support the theatre of operations with a considerable degree of flexibility. It had tandem ejection seats, retractable landing gear and was equipped with a 750shp Pratt & Whitney Canada (P&WC) PT6 turboprop. In addition to being reliable and in operation worldwide, Embraer had already mastered the handling of the PT6 engine, which was in use on the Bandeirante.

Under the wings, the aircraft had six hardpoints and its design was bolder, offering improved range and manoeuvrability. The aircraft employed an all-metal semi-monocoque structure, with wings built in three main parts, a fuselage in two parts and a canopy integral with the windshield. Internally, each wing had two 215-litre (56.7-gal) main tanks, a 210-litre (55.5 gal) leading-edge tank plus two 100-litre (26.4-gal) auxiliary tanks, totalling 525 litres (138 gal) in each wing and 1,050 litres (277 gal) in both. Externally it could still carry two 330-litre (87-gal) jettisonable fuel tanks, adding 660 litres (174 gal) and totaling 1,710 litres (451 gal) including the internal tanks.

As for the underwing pylons, these allowed for the loading of various general-purpose bombs, machine gun pods, reconnaissance pods and rocket launchers – four were capable of carrying loads of up to 340kg (750lb) and two up to 295kg (551lb) – totalling 1,950kg (4,299lb) of ordnance. In a configuration with two Mk 82 (500lb, 227kg) and two Mk 81 (250lb, 119kg) low-drag, general-purpose bombs, the combat radius would be 984km (329nm) and the aircraft would require a runway of only 1,500m (4,921ft) .

Comparisons were made with the same models selected to evaluate the EMB-301 and, according to schedule, the inaugural flight would be made within 18 months of the launch of the programme.

In 1977, Embraer submitted projects EP-301/105 and EP-311/104 to the FAB. The Brazilian manufacturer already saw the possibility of selling the two projects abroad as well, and that year presented them at the Paris Air Salon at Le Bourget in France.

'Guido was thinking a lot about the EMB-120 Brasilia. At that moment he wanted the project for this trainer to be as simple as possible, preferably using only a turbo-

prop engine like Pilatus did with the PC-7. He did not want another programme; he was asking me to make a version of the T-25 Universal with a turboprop. It would keep me and Neiva busy and would not compromise the EMB-120 Brasilia', Kovács later revealed.

The Aeronautics General Staff finally showed interest in the projects and several meetings were held with FAB officers with the objective of gathering opinions, suggestions and criticisms. It was obvious to the Embraer team that it would be necessary to abandon the EMB-301 and make several modifications to the EMB-311.

Enter the EMB-312

Without wasting any time, Embraer sent its project EP-312/130 for an advanced training aircraft capable of performing aerobatic manoeuvres – the third configuration – to the FAB in August 1977.

Among the changes in this design were the removal of the ejection seats, reduction of overall length by 15cm (5.9in) and relocation of the instructor's seat to a higher position so that he would have better vision over the head of the student.

In his book *The Takeoff of a Dream* (Lemos Editorial, 1999), Ozires Silva observed: *'We started an analysis including practically all types of military training aircraft in production and operating around the world. It was not difficult to understand that, although all the air forces had embarked on military training programmes with jet aircraft, we thought there would be room and even a very good future for propeller airplanes. I particularly insisted on that – and recalled my course at the School of Aeronautics; I was an ardent advocate of propeller aircraft as more effective equipment for training young pilots. On the other hand, it was very clear that these airplanes could offer better value for money. They were slower, more manoeuvrable, consumed less fuel and, we understood, they presented flight situations that better trained the students.*

'We spent much of our time generating alternatives and exploring ideas in detail to try to produce something new in the field of instruction and training of military pilots. There was the need to devise something innovative, but at the same time at costs that could be covered by the ministry's tight budget. We could not forget Lélio Lobo's remarks, which cast doubt on Embraer's ability: the time it took to create new aircraft and certainly the volume of contracts that the Ministry of Aeronautics already had underway with Embraer.'

Priority missions continued to be those of advanced training including aerobatics, training, day or night and IFR flying and using weapons if necessary, for the instruction of the military pilot. Secondary missions encompassed special operations, close air support, reconnaissance, armed escort and attack. Resistance, durability, simplified maintenance, low operating costs, high availability rates and limited dependence on ground infrastructure to fulfill the assigned missions were also part of the FAB's requirements. Flexibility was also paramount to allow the aircraft'd use from remote bases and on unprepared runways. For the crew, the aircraft should provide good visibility and be manoeuvrable.

The EMB-312 and its evolution

The powerplant chosen was the 680hp Pratt & Whitney PT6A-27. With it the aircraft would be able to carry up to 1,480kg (3,262lb) of stores on six hardpoints under the wings, attaining a speed of 397km/h (214kts) at an altitude of 7,620m (25,000ft) or a maximum speed of 422km/h (227kts) at 4,572m (15,000ft). Three fuel tanks in each wing – two main tanks with 215 litres (56.7 gal) and one in the leading edge holding 185 litres (48.8 gal) – provided for a total of 800 litres (211 gal) of fuel. With this fuel load it was possible to carry almost 900kg (1,984lb) of ordnance over a combat radius of 1,100km (593nm).

But despite the numerous and intense meetings, Embraer's philosophy and understanding of an advanced training and ground-attack aircraft was not the same as that of the FAB. This led to disagreements between the two parties. The FAB still envisaged a jet aircraft, a concept that already enjoyed international success and which was serving in many air forces. But the price of fuel and the maintenance of jet engines would be a major impediment. Eventually, it was clear that the final decision would fall on a turboprop aircraft.

On the other hand, the configuration of the seats remained somewhat controversial. The FAB had clearly became accustomed to having an instructor and student sitting side by side. This made teaching easier because the instructor was able to see the student's mistakes and immediately demonstrate, in practical terms, the correct control or power input. In the same sense, the student could more quickly and easily absorb the lessons from his instructor. But Embraer believed that the tandem configuration could be the right one. With it, the student would be brought even closer to a fighter-like environment without having the instructor at his side.

The EMB-312 was the first turboprop aircraft in the world with a single lever to control engine power. The solution was latter adopted by Pilatus on its PC-9.
(João Paulo Moralez)

But while this topic was still open for discussion, the use of ejection seats was not. When discussing the proposal with the FAB, and above all with the instructors at the AFA, the installation of ejection seats for the crew was specifically requested. This was the first major challenge facing the project. Despite having experience with ejection seats, obtained with the construction of the EMB-326 Xavante, this would be the first time that Embraer had designed an aircraft using this system. In terms of weight it would be equivalent to the aircraft carrying four people and the cost would represent 30 per cent of the final price – around USD 400,000. The idea was that the ejection seats would be the same as those in the Xavante, namely, the Martin-Baker Mk 4 BR04B.

Outside the FAB, ejection seats would be an optional item, to be chosen by the particular customers in other countries. But the ejection seats, along with other Tucano-specific items, would ultimately be responsible for the success of the aircraft on the international market. In fact, no Tucano was ever sold without ejection seats fitted.

But even if it was offered with ejection seats, this alone was not enough for Embraer to enter a market as competitive as this one, and to go up against traditional suppliers like Beechcraft and Pilatus. The manufacturer needed to add values of modernity and innovation to its aircraft. This would be fundamental to continue building the company's presence in the foreign market – where Embraer would sign most of its civil and military contracts.

And the Tucano itself, because it was a turboprop aircraft, needed to show that it was on par or even a level above its jet-propelled rivals – and not convey an image of inferiority.

There was something else different about the Tucano project.

Ary Guilherme Leber was an Embraer engineer who worked on the Tucano:

'At the time, everything that came from abroad was better than what was made in Brazil. There was this mindset here and nothing was national. The great secret of the people involved with the Tucano, including me, was that we were a group determined to do things here and shows that Brazil was competent. I went to Embraer to make airplanes and Guido was also a person who wanted to do things in that way. Many people said that he was a visionary, that it was not practical. What really matters is that he wanted to develop technology in Brazil.'

Embraer had a tough problem to manage. The aircraft had to be simple and with low costs of acquisition and operation. But it would also have to bring together features that demonstrated an image of all-round modernity.

While Kovács had a mindset in which everything was conceived in a way that was as simple as possible, Pessotti constantly flirted with modernity. Simplicity impacted positively on cost, weight and strength of the airframe, and the development deadline. Modernity could adversely affect all these points.

Striking the balance between these two worlds, and acting with thoughtfulness, was not an easy task. But the rivalry that existed between Kovács and Pessotti, always kept within healthy limits, ended up largely benefiting the Tucano project – which was both simple and modern.

From its conception, the EMB-312 was projected with four underwing stations to carry up to a tonne of conventional weapons including bombs, rockets and machine guns. (Brazilian Air Force)

As far as the pilot was concerned, the cockpit concept and the Tucano's way of flying resembled those of a jet – but without the restrictions and costs involved. The sight, instrument panel and rearview mirrors were arranged in an architecture reminiscent of a fighter. The instruments on the panel, their presentation and composition were intended to give that impression to the pilot. But that was not enough.

The Tucano featured other novelties. The engine would be started by pressing only a single button.

Another highlight was in respect to the engine power and propeller pitch control levers. On aircraft such as the PC-7 and T-34 Mentor/T-34C Turbo Mentor the levers are separated and the pilot has to command each in an isolated manner.

Embraer gave itself the challenge of creating a control unit with only one lever, as found on fighters. Thus, the pilot would only have to concern themselves with applying or reducing the power; the lever's control box would make all the necessary compensations in terms of propeller pitch and injection of air and fuel.

'I called the Pratt & Whitney Canada agent who was at Embraer and asked if we could create a mechanical programme so that we could make the power and propeller levers together in one. This agent said that our idea would not be possible and I questioned why he had such a quick response and if that was his final word. We had a great relationship with them and the agent said he would contact the company headquarters in Canada. The president of Pratt & Whitney Canada called me saying that what we wanted to do was very complex and I argued that Embraer could do it. We were their biggest customer worldwide and in the end they did not refuse our project', explained Ozires Silva.

By controlling the propeller and power levers at the same time, the pilot would not have to look at the instruments to determine the propeller's rotation in order to achieve the desired power. This brought the Tucano closer to a jet.

But even so, one point remained open. In a propeller aircraft, each time the pilot increases power, it is necessary to correct the inertia generated in the fuselage by using pedals, commanding the rudder in order to prevent the aircraft from drifting sideways.

'In the North American T-6 we flew at the Aeronautics School, the engine developed 600hp and the airplane was relatively lightweight for that engine power. Then there was a very strong propeller torque, which was corrected from the aerodynamic point of view. And I was saying that aerodynamic correction had nothing to do with torque. The idea I proposed to the team was to add a deflection on the engine, providing a side force component to compensate torque. This deflection, 3° to the left of the aircraft's longitudinal axis, virtually eliminated torque effect', Ozires added.

In fact, this characteristic was noticed and highly praised by all who flew the Tucano on visits to Embraer or on exchange with the Brazilian Air Force. When the Royal Air Force was studying the possibility of incorporating the Tucano into the training course for its military pilots, one of the British instructors testing the Tucano remarked in amazement that he did not even feel the torque of the propeller: *'This plane feels like a jet!'*

Back in the struggle between modernity and simplicity, there were still some other suggestions.

'We used structural bonding to attach the tailfin to the rear fuselage. Could we use rivets? Yes, we could, but Guido wanted Embraer to consolidate the use of structural bonding. We learned to handle and work with composite materials, from which the

main landing gear doors were made of, as well as using chemical machining on the wing and fuselage skin. In this case, we protected a thick plate with a kind of mask at the points that needed to remain thick, therefore more resistant. The parts that would not undergo so much structural stress in flight did not have to be that thick, so we would use a chemical bath to make them thinner through controlled corrosion. This greatly reduced the final weight of the aircraft. All this requires an exact calculation of the chemical mixture, the time that it will remain in the bath, the thickness of the plate, to work well and not miss the measurement. The rivets were not exposed and the plate joints did not overlap one another. We even considered the possibility of puttying the surface and polishing it to make it smooth, like that of a glider', commented Leber.

Another point of great discussion, again aimed at modernity, was in the retractable entry step, which would close when the landing gear retracted. It was decided to develop this system – complex and subject to a greater number of failures, according to Kóvacs – to reflect the modernity of the aircraft. Kovács had proposed a fixed low-drag step, but his idea was rejected.

'We already had nine external antennas – some with a drag magnitude comparable to a well-structured step. The sensible compromise favoured a discreet fixed step. But while an antenna is a symbol of modernity, the fixed step is one of anachronism', commented Kovács.

But while it was necessary to innovate, Embraer also had to save time and money. In the case of the Tucano programme, both were scarce.

By the third configuration, which was introduced in 1977, Embraer chose to use solutions and even assemblies and parts of other projects in production to reduce the project's risk and costs, such as the rear fuselage section of the T-25, for example. Again, while time and a cost reduction might be gained, there was a knock-on effect on development, since specifications needed to be developed by adapting something originally created for a completely different purpose.

'Some people were against copying or taking advantage of anything that came from Neiva. It would be just Embraer, everything brand new. We do have a tendency to make things easier, move forward taking advantage of what is ready – and this does not always work.

'Gradually the subassemblies and parts from other projects gave way to new designs. The extensive use initially recommended was rejected, although retaining a main landing gear inspired by the Navajo's. Because it was bulky, it had to be housed inside the wing. I only insisted on advocating a new aileron, with the wing-tip cut, more suitable for aerobatics, as well as a change in the flaps and in wing structure', commented Kovács.

This change of mentality began in July 1978, when major changes were made in the project drawings.

In the fourth configuration the Tucano gained an aspect of greater modernity, including a swept-back tail. The landing gear remained similar to that of the Navajo, with the distance between the oleos remaining narrow to accommodate the wheels inside the fuselage – allowing for thinner wings. But this reduction in the distance between the landing gear legs would also reduce stability during ground manoeuvres.

Thus, the fifth and sixth configurations had an enlarged distance between the oleos, which resulted in the need to house the wheels inside the wings – increasing their thickness.

Changes were also made to the nose landing gear doors. The initial suggestion was that the wheel would remain partially exposed with the landing gear in the retracted position. Another option was the development of a large side door that would remain open when the aircraft was on the ground, or sliding doors of the telescopic type for the nose landing gear, which would have a simpler mechanism. However, the Technical Board opted for smaller side doors, divided into two parts on each side. All doors would be open as the landing gear was lowered. On the ground the two doors to the rear would remain closed, opening only to allow the operation of the landing gear.

Once the flight test campaign started and the aircraft's operation was observed, the system was found to be more complex and difficult to adjust, and in flight the doors would usually be half-open, due to the aerodynamic loads experienced during the manoeuvres.

The internal tanks were modified as well, but would be made of rubber and divided. Engineers proposed the development of integral tanks instead of rubber tanks; however, the Technical Board rejected that option on 5 September 1978. They would later reverse their decision.

The main spar underwent changes, as did the wing root, which became larger. But there would still be other major modifications that would lead to the design's seventh configuration.

Embraer was meanwhile struggling to deal with the loss of Major José Mariotto Ferreira, who was killed testing a spin manoeuvre with an Aerotec T-23 Uirapurú. An experienced pilot, he was part of the crew that made the Bandeirante's maiden flight. He was a friend of Ozires Silva and other members of Embraer's board of directors, having been invited several times to join Embraer permanently, dedicating himself exclusively to the manufacturer and leaving CTA. But he chose to continue serving the two institutions. On 1 November 1968, days after the inaugural flight of the Bandeirante prototype, tragedy struck.

'One of the things that gave us much trouble were the spin manoeuvres. At the time, just before that, by chance and unfortunately, we had a terrible accident with Major José Mariotto Ferreira, who was lost testing a spin manoeuvre with the Aerotec T-23 Uirapurú prototype. A few days before the accident I talked to him, and he told me that he did not like parachutes, that he did not even think about them. And in the Flying Officer course he did at AFA there was no instruction for an emergency jump with parachute. But the Uirapurú had spin problems and we thought he would have trouble at this stage of flight.

'With Major Mariotto, the airplane went into a spin and it triggered the spin recovery parachute, which still did not prevent the aircraft from leaving the manoeuvre. He felt the problem and must have bailed out of the airplane right away. However, as Major Mariotto wasn't familiar with parachutes, he ended up opening it too soon, got tangled in the T-23's tail, falling along with it. After the accident, Brazil contacted France and tested a model of the T-23 in the wind tunnel of the Office National d'Études et de Recherches Aérospatiale, in Lille, exactly in the spin. After several tests, it was determined that installation of a fin in the lower part of the tail

would prevent the aircraft entering a spin and not lead to this fatal manoeuvre. And that problem was solved', explained Kovács.

It was natural for Ozires Silva and for everyone else to request that the same test be made with the Tucano and that the new aircraft should have excellent flying qualities, above all in aerobatic manoeuvres. What had happened with the T-23 should not repeat itself with the Tucano.

Kovács was against the use of a bolder, more modern tailfin. For him, that type of fin would not favour spin manoeuvres and the result was likely to repeat the situation that occurred with Mariotto. Kovács proposed a more conventional and elongated tail, but the reluctant Technical Board at first did not want to abandon the swept-back tail, requesting that studies about its stability be made. Meanwhile, the aircraft's design was intended to be suitable for both types of tail. The main change of the seventh configuration was thus defined. Having adopted a more conventional and classic tail, the use of fins in the rear section of the fuselage was dispensed with.

From the start of the development phase it was intended that no wind tunnel tests would take place. There were several reasons that led the team to take this decision. There would be no time to build a model, much less money to do it. Another problem was using the CTA's own wind tunnel, which tended to be bureaucratic and time-consuming.

The alternative decision fell on an unprecedented solution – at least in Brazil – involving the construction of a radio-controlled Tucano flying scale model – already incorporating the conventional tail of the seventh configuration – so that flight characteristics could be verified. Several tests were conducted but the model crashed during the spin tests, after the model suffered interferences in the radio transmission.

After tests with the model and engineering studies, the Technical Board opted for the conventional tail.

Embraer completed two cockpit mock-ups to test ejection seats. One tested the detonation of the canopy, and the second, mounted on a rail, the ejection procedure itself. Here, the canopy is being prepared for an emergency opening test at São José dos Campos.
(Embraer, Archive Ary Guilherme Leber)

The final configuration

Although almost the whole project was now defined, there would still be some more fundamental modifications so that construction of the prototype could be completed. The changes introduced in 1979 would be significant.

The first addressed the Tucano's engine, replacing the 660hp PT6A-27 with the 750hp PT6A-34. Finally, however, the choice fell on the PT6A-25C, also developing 750hp, but with a lubrication kit for high-performance aerobatic manoeuvres, especially for inverted flight.

'Embraer learned that another aircraft, very similar to the EMB-312, manufactured by Pilatus in Switzerland, would leave the factory with a 950hp turboprop engine – the PC-9. The EMB-312 was already being prepared to fly with a 680hp engine, enough for that category. Until then, it was known that they had asked Pratt & Whitney to modify the traditional engine to be aerobatic, that is, there would be the need to collect oil for the pump and continue with the engine oil breather also in inverted and knife-edge flight. At that time there was no knowledge of any type of training aircraft flying with an aerobatic turboprop engine; we believe the Swiss utilised some ideas from Brazil.

'Not to be left behind by the competition, Embraer changed the power of the EMB-312 to 750hp and, in practice, FAB gained a very 'ferocious' engine, because if it was already going with a well-sized power unit, now it would be oversized. The PT6A-25C operates at a maximum torque of 1,795lb [814kg] *on the propeller shaft, but we consider it a very high and unnecessary effort when using it in full. After its operation began we requested a test flight and that the minimum power required for take-off should be applied. Temperature that afternoon was 32°C* [89.6°F] *and our runway in Pirassununga is at the 2,000ft level* [627 MASL]. *The Tucano took off with 1,300lb* [590kg] *of torque, demonstrating that it can save a lot of power and fuel, in addition to providing a longer engine life'*, explained Nelson Marcomini, then a sergeant and aircraft engine mechanic.

With a more powerful engine, the tailfin was increased in height by 15cm (5.90in). The next change was the elimination of the wing leading edge extension, at the root. Finally, another change addressed the canopy.

The idea of using a windshield was abandoned and in its place would be adopted a side-opening, frameless single piece. This would be made of acrylic. This could be produced using a mould, to ensure the correct shape and not affect the aerodynamics. However, when the instructor in the rear cockpit looked forward, a moulded transparency would lead to optical distortion – impairing his vision. To meet impact specifications and remove optical distortion, the part had to be vacuum-formed by pressure and without a mould, like making a bubble. The process required special kilns as well as more costly and time-consuming procedures. To find a solution, an Aermacchi engineer was called in to assist Embraer.

The idea was to provide ample vision, as in a fighter. In the end, the canopy weighed around 50kg (110lb) including all its structure, had a constant thickness of 8mm (0.31in) and could withstand medium-sized birds at a speed of 250km/h (135kts; corresponding to landing and take-off phases, when the risk of a bird strike is greater). Due to the increased weight it was necessary to adopt an explosive device to jettison the canopy in emergency situations on the ground.

The first prototype YT-27 1300 configured with six SBAT 127mm (8.5in) rockets. The aircraft is in its final configuration, with dorsal fin at the root of the tailfin, and fins at the root of the horizontal stabilisers. (Embraer)

The downsides of this manufacturing process were the longer time required to replace the transparent canopy element if necessary, in addition to reducing its resistance against bird impacts – however, its tolerances were still within the required specifications.

'We developed the canopy according to Guido's idea, with conical plane profiles which provided optical quality primarily to the instructor in the rear position, who would not experience any deflection in vision when looking directly ahead through the canopy. This also increased resistance because the canopy would not go through deformations over the years, since internal tensions were much reduced. The item also did not experience the crystallising effect', Ozires explained.

'Tests were made at Martin-Baker in the UK to find out which would be the best option when ejecting the seats, i.e. ejecting the full canopy assembly or breaking it with a detonation chord. It was found in the tests that none of the options was feasible, as shrapnel would hit the crew on board.

'The solution was to break the canopy using the seat itself, through the acrylic. To protect the instructor in the rear position, an internal windshield was provided, something that did not exist in the initial project', Leber added.

This was now known as the eighth configuration.

'For the airplane to have excellent characteristics in a spin it is necessary that the nose is as short as possible. The longer it is, the worse its features in these spin manoeuvres. But turboprop engines are lighter and longer, thus putting the

propeller further ahead, far from the centre of gravity. To make matters worse it had been defined that the airplane would have the ejection seats with a difference in height between them, under an integral canopy without a windshield. The seats have a clearly defined trajectory and all these characteristics of the design combined required an ever-increasing length. This gave us a lot of work and the solution was to leave the nose as short as possible by gathering the components in the cockpit region and putting ballast in the nose as compensation', commented Kovács.

Development

Parallel to the evolution in the development and gradual improvement of the various Tucano configurations, Embraer was still in negotiations with the FAB to finance the programme. The air force was confident in the project, so much so that in 1978 the EMB-312 received the military designation T-27.

But even if they were enthusiastic, the then Ministry of Aeronautics had already invested heavily in programmes like the AT-26 Xavante and the C-95 Bandeirante.

Brigadier Joelmir Campos de Araripe Macedo, then Minister of Aeronautics, claimed that the ministry did not have funds to finance Tucano development, but there was the possibility of an order to fill the FAB's trainer shortfall if the turboprop was already available. It was then that Embraer opened discussions with the Financiadora de Estudos e Projetos (FINEP, Funder of Studies and Projects), an institution that had just been created and was directly linked to the Ministry of Planning. Finep aimed to place public funds into business risk to invest in new projects.

'I was in contact with Finep and confess I did not have much hope of achieving two things: approval of the project, because of the refusal we received from Brigadier Araripe, and that the money would be enough. At first, I was wrong and I am sure that Finep went over and approved the funding thanks to the interference of Brigadier Menezes and Brigadier Nelson de Souza Taveira, as they collabourated a lot in this regard. Like us, they were convinced that Brazil should enter this field. Brigadier Menezes in particular was crucial, as he helped to draw attention to what we were doing in São José dos Campos. Even at the beginning of Embraer few people believed that we could flourish. In the second instance I was right: the money was totally insufficient. It was something symbolic; it wasn't enough for tooling', explained Ozires Silva.

In September 1978 an agreement with Finep was signed for a sum of USD 150,000.

'The T-27 was the first airplane to receive funds from Finep, for development and production of the prototype. At this time I was already director of the Aerospace Technical Centre, now the Department of Aerospace Science and Technology. Accompanied by Brigadier Taveira, I went to the agency to get the money to fund the Tucano project. The aeronautics development programme at Finep was coordinated by Brigadier Hélio Langsch Keller, who was my commander and a veteran of the 1st Fighter Aviation Group in Italy during World War II. He listened to us and immediately allocated resources to CTA, which would be destined for the development of Tucano', concluded Brigadier Menezes.

On 6 December 1978, shortly after the funding signature with Finep, Embraer signed a contract with the Ministry of Aeronautics for the development of four prototypes, two for in-flight trials and two that would be used in ground-based structural tests.

The workflow of the programme, however, would only begin on 15 January 1979, when the teams were defined and structured. In that sense there would be one more great difficulty to overcome – the lack of human resources. Many engineers were already engaged in other programmes already under way, or in studies for new projects, such as the EMB-120 Brasilia.

With the project approved by the Technical Board, the engineers began to design the various systems, components, structures and other parts as the project quickly gained shape.

While discussing the best engineering solutions towards the final configuration, the eighth, Embraer created four mock-ups. The first, for engineering, was used to install the various electrical, electronic, hydraulic and air-conditioning systems, among others – to check if there was any conflict or interference between them. There was no outer skin, only an internal structure. The second mock-up was restricted to a cockpit for test pilots and AFA instructors to gather opinions and suggestions from the military that were analysed. Many of them were implemented later by Embraer. For the most part the changes were in relation to ergonomics and the positioning of certain systems and consoles. There was also a third mock-up in which the canopy detonation tests

Part of the team that built the first Tucano prototype. In the coloured shirt is Joseph Kovács and to his left is Ary Guilherme Leber.
(Embraer, Archive Ary Guilherme Leber)

were made for ejection studies. A fourth mock-up of the cabin mounted on a rail was shipped to Martin-Baker in the UK for rehearsals of the ejection itself.

Since Tucano also foresaw international sales, the mock-ups presented a perfect opportunity for Embraer to use as a means of publicity and marketing.

'There have been unremitting restatements of corrections in cockpit layout in opposite directions. This happened according to whoever evaluated it, depending on height and hierarchical position. Often people came with exaggerated height and physical conditions, both on one extreme and the other. In the end we got a cockpit more spacious and less limited than the Xavante's.

'The seat chosen was the Martin-Baker Mk 8L, positioned at a 17° angle to the aircraft floor and the cockpit layout allowed people with a minimum of 1.60m [5 ft 3in] *in height and a maximum of 1.90m* [6 ft 3in] *to fly the Tucano. This allows a total of 98.75 per cent of military pilots to be able to fly the aircraft. It is the smallest limitation that can be achieved'*, commented Kovács.

The Tucano project, to a certain extent, introduced new procedures in the development of Embraer's aircraft. The manufacturer had a dedicated computer for engineering calculations, aerodynamics and planning. The rest was done on the slide ruler and drawing board with China ink.

'Kóvacs coordinated the construction of the airplane in the project area. He was much more on draft, with his own team, but he did not draw himself, he specified. He was the programme's general manager and followed and directed what happened in each pre-project area – which defined the airplane itself. And unlike other airplanes that Embraer produced, the Tucano was more sensitive and its construction had to be very well done. Kovács wanted something very close to a glider, with high yield, including rivets that were not exposed. Being very refined, any variation was felt in flight. Embraer had manufactured the Xavante, which I worked on, but they did not know how to make an airplane as sophisticated as the Tucano', explained Leber.

The designers drew the pieces of the large subassemblies, detailing what was needed, so the time was much better optimised. From the outset the timeline stipulated dates for all processes, since time was scarce and control had to be strict. The normal forecast was for such a prototype to be ready in two years, but the team would only have six months for this endeavour.

There was only 180 days between the first component built and the first flight. Leber was the person who made the link between production and engineering; he was the coordinator between the design and manufacturing sides. He had to know requirements for the design of a particular piece needed for production, then go to the design area and get the drawings straight from the designers' drawing boards – sometimes there were dozens at a time. The project teams worked on everything at the same time, and he managed the project so no one would stand still.

In some cases, he took engineering suggestions from production and introduced design modifications, so that a particular part would be easier or, in some cases, viable to be built.

'I went through several areas of Embraer production between 1974 and April 1978. In November 1979, after a year and a half off, I had already returned to work on the EMB-312 project. The aircraft was built within a timeline with dates for everything because we had to make the airplane in six months exactly.

'I was the person who needed to know the part required in production to get the design on the project, working in advance to avoid delays. Sometimes I would take the drawing with the ink still wet, send it to be photographed and then pass it to manufacturing. We also sought to put solutions within Engineering itself, for things that would not be possible to be manufactured in that way', Leber added.

One of the most striking moments in the development of the first prototype, YT-27 FAB 1300, was the beginning of fuselage construction on 2 February 1980. Ten days later, at exactly 18.00, the manufacturing teams began working on the right main spar. *'The spar was a 5.5m long* [18ft] *machined 'T' profile, starting wide at one end and tapering until it reached the other end. This profile had an angle of 5° and would come ready from the manufacturer; however, due to problems in the supplier, there would be a delay in receiving the spar. The part had no welds and needed to arrive perfect. We had the 'T' profile at the factory, but with a 90° angle. In order not to delay production, we tried right there in the company to shape that 'T' profile in the way we needed, with thousandths of a millimeter precision. The solution was to put it in the hydraulic press, with a block of steel, to reach that measure. It took a meeting with Engineering to get that approval, because getting the right angle on a machined, cold folded and tempered piece of steel was very difficult. And we did it on the second try. Engineering did the maths and approved the job.'*

But it was still necessary to machine this piece, which began thicker at one end and tapered towards the other end. This tapering was completed in the form of steps. But at that moment the team did not have the machine or the computer with the exact outline to be followed in the machining. The solution was to use equipment that checked the dimensions of the parts and the precision of their manufacture in the micron scale, which corresponds to a millionth of a millimetre. This system was instrumental in ensuring the standardisation and quality of the aircraft manufacturing process.

But, instead of inspecting it, the team reversed the process. The inspection machine had the outline to check the precision on the spar when it was ready. Thus, the spar was painted with hydrographic paint and the machine scraped it in the place where the piece was to be machined. The entire process was successfully completed and the part was approved by Engineering.

Work on the left main spar was begun at 12.00 on 23 February and both components were completed on 4 March, each weighing 24kg (53lb).

Assembly of the wings gave a new boost to the project, which was already taking shape in Hangar F-60. The teams began work at 06.00 on 7 March 1980, concluding placement of the final rivets at 16.00 on 2 May. Two days later, the fuselage was finished and the job of joining it with the wings began on 25 May, taking two days to complete.

'The airplane is very compact. We had to follow the standards separating the hydraulic system from the electric. The most difficult part of the Tucano was the front fuselage, because it has a lot of parts. Engineering showed the front fuselage ready on paper, but called Production to say that they did not know if it would be possible to mount it, because it was too complex. After much talking we concluded that to do that would require a dwarf with long arms and small hands to reach certain places. It came out just like that; we just changed some rivet positions', explained Leber.

The elevators were the last large structures to be placed on the EMB-312. From 15 July, the aircraft began to receive the last details in preparation for flight.

'The inspection windows were made with the aircraft finished. The Engineering people went down to Production and then we started to open the windows where it was necessary', Leber explained.

The test pilot chosen for the first flight was Colonel Luiz Fernando Cabral, who had served with the FAB. He would occupy the front seat of the aircraft while the in-flight test engineer, Gilberto Ideo Otaka, took the rear seat.

With the Tucano prepared for its first flight, the teams had begun the first engine tests on the ground, when an inadvertent opening of the canopy occurred after the power of the PT6A-25C was raised to maximum. The problem occurred because the lock was not correctly regulated, a problem that was solved immediately.

The next steps were taxi trials and take-off runs during which the stability of the aircraft was checked.

Everything seemed to be ready.

The EMB-312 carries 445kg (981lb) of systems and equipment, excluding ejection seats. This comprises 19 per cent of the maximum take-off weight. Apart from pressurisation, the new aircraft had everything that might be found on a subsonic jet or a twin-engine trainer. There had been over 4,500 drafts and drawings. In the course of preliminary design, development and construction of the prototype, 250,000 hours had been spent on engineering, 113,000 on tooling and 50,000 for the construction.

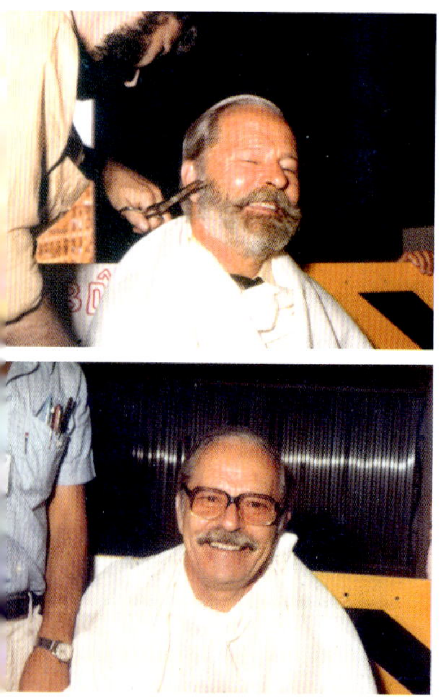

After winning the bet, Ary Guilherme Leber starts to cut Kovács' beard with a pair of scissors.
(Embraer, archive Ary Guilherme Leber)

Kovács' beard and the first flight

On 23 April 1979, it was decided that the inaugural flight of the EMB-312 would be made on 19 August 1980, the date of Embraer's 11th anniversary.

But among the board, design team and production team there was great doubt as to whether the EMB-312 would fly on schedule.

'Once a week, we would meet Kovács and give him updates on what was happening', Leber recalled.

'Closing the meeting, Kovács said the airplane would not be ready in time. Some said it would be, others not. 'So I propose a bet. A box of imported whiskey on the airplane not getting ready', *he said. Domingos Regino, head of manufacturing, shouted,* 'That's not much! I bet two boxes! One imported and one national'. *Something that had been in my head for a long time was Kovács' beard.'*

The designer of the EMB-312 had grown his beard after the death of his eldest son, Marcos, in a glider accident. He had suggested his father grow a beard and Kovács decided to attend to his son's request after this tragedy. Anyway, at this point, the bet favoured the smart Hungarian engineer.

'A new child was being born for him; it was the Tucano. Then I said: 'I bet my beard against yours that the Tucano will fly before August 19! And you're going to have to shave your beard in the middle of the F-60, in front of all the staff. And you will lose!''

The bet was made.

Five-thousand spectators came to witness the first public flight of YT-27 1300 on 19 August 1980.
(Embraer)

16 August 1980

In front of the Embraer employees, an unfamiliar roar echoed on the taxiway as the Tucano's engine was started. Several ground tests had already been made, including a number of take-off runs. Manned by test pilot Luiz Fernando Cabral and engineer Gilberto Ideo Otaka, the EMB-312 moved onto the runway, lined up at the head of the track, applied maximum power and took off on its inaugural flight. The success of the new Embraer trainer was sealed. Leber, as well as many other employees, boke down in tears as he watched the evolutions of the first EMB-312 in the skies of São José dos Campos. Serial number 312001 received the registration number 1300, and in its YT-27 military designation the letter 'Y' indicated that it was a prototype. 'T' signified a trainer.

'I was beside Kovács, I cried a lot. I slept a few hours a day, I did not have weekends, we arrived at the company at 06.00. There was a loaf of bread and coffee, there was no time to leave, and I stayed in the factory for 48 hours without stopping. After the airplane had made the first flight, I did not need that madness anymore, and I worked eight hours a day. I did not even know what to do in my free time', recalled Leber. However, this milestone was not yet the official first flight for the authorities.

First flight, first fright. After performing some manoeuvres to check the basic controllability and stability of the EMB-312, an alert in the cabin left everyone worried. Apparently, the nose landing gear was not locked. For more than two hours the aircraft was in orbit over São José dos Campos, while the crew manoeuvred the aircraft, trying to lock the landing gear in position before all the fuel was consumed. If an emergency landing was needed, the chances of fire would be reduced with minimal fuel in the tanks.

Unable to lock the landing gear, the crew had to attempt a landing. Fortunately, it was a success. It was pure luck. The gear locked when the tyres touched the ground.

With the requirement to prepare the aircraft for its official flight, which would be made three days later, there was no time to discover the reason why the lock had failed on the nose landing gear. As a precaution, the doors of the front gear were removed for the official flight on 19 August.

Two reasons to celebrate

Embraer was celebrating. It was the company's 11th anniversary. Five-thousand people had gathered in the aircraft's parking area, including government officials, officers and armed forces representatives, suppliers and others. The event included CTA's band, a podium for speeches, and everything else a ceremony for a first flight deserved.

The ceremony began at 15.00 with the arrival of Brigadier Victor Didrich Leig, representing the Minister of Aeronautics Délio Jardim de Mattos, and his entire entourage. With a bottle of champagne, Embraer employees José Benedito Guimarães and Sérgio Landi baptised the YT-27 1300 prototype.

The National Anthem played as the ceremonial duties were fulfilled to the letter.

At 15.30 Colonel Cabral pressed the start button and the powerful turboprop began to roar. While he checked the parameters of the engine, oil pressure, temperature, operating systems and flight commands, eight T-25 Universals from the Air Force Academy flew over the São José dos Campos airport.

The EMB-312 began taxiing while the pilots checked the instruments. Aligned at the head of the track, the power lever was pushed forward.

At last, the take-off run and at 15.35 the EMB-312 took to the air.

After the Bandeirante, Ipanema, Urupema and Xingu, the time had come for the Tucano.

The pilot's voices were heard by everyone on the ground. The aircraft reached a speed of 333km/h (179kts), 80km/h (43 kts) below the specified speed. Some curves were flown, with greater and lesser intensity.

After 20 minutes, the YT-27 returned, escorted by a T-25. All nine aircraft passed over the public and then landed.

'On a first flight, the envelope of operations is very limited and one needs to feel the aircraft's reactions and responses to the commands. However, we could see that the T-27 has excellent flight characteristics, performance, and stability, important qualities for a training aircraft. Now we are going to continue a progressive programme in order to know the aircraft area by area and make the adjustments that are always necessary in a prototype. This adjustment phase should last about three months', said Colonel Cabral.

According to engineer Otaka: *'in the first flight we basically verified the stability and control of the aircraft and we can affirm that all the answers were positive, corresponding to our expectations. The aerodynamics of the airplane are perfectly suited to engine power. Our testing programme follows the FAR Part 23 standard and is based on several steps; in 90 days we intend to fully evaluate the aircraft in terms of stability and control items. The second prototype, due to be ready in December, will test systems specifically'.*

Unable to attend, Brigadier Délio Jardim de Mattos telegraphed: *'Coronel Ozires Silva – Embraer SBSJ – at the time of the flight of the prototype T-27, transmit directors board, as well as all employees of Embraer, expression of my pride in finding that the trust deposited by us has been rewarded with honesty and work. This victory is an eloquent example that success depends not only on ability but rather on determination without limits and a clear vision of national reality. The success of Embraer rests on the realism of those who lead it, men capable of understanding*

The first and second prototype fly together, still in the original configuration without the dorsal fin at the root of the horizontal stabiliser.
(Embraer)

that a company does not live from utopian dreams, but from the results of its balance sheets. Sincere thanks.

Délio Jardim de Mattos.'

But in the midst of so much excitement and celebration, there was a loser for the bet made weeks earlier.

'After the flight an 'Order of Manufacture' was issued so that Kovács' beard was shaved in Hangar F-60. The Production Control staff went to pick him up in his room and on a stage set in the middle of the crowded hangar I made the first cut with a pair of scissor shears, which symbolised those who had put a lot of effort into making the flight a success. Finally Quality Control stamped the 'approved' on his chin. The bet was won', Leber remembered.

Trials campaign

After the inaugural flight of the first prototype, the draft team presented 255 modifications, divided into four groups, to be implemented in the project. Most of them were relatively minor refinements.

Since the low-altitude aerobatic manoeuvres conducted on the third flight of the aircraft had not presented any serious problem, it was decided that only the necessary modifications – 11 in all, would be made. The rest were postponed.

The project team immediately began working on a new balancing mechanism to facilitate the opening and closing of the heavy canopy; the installation of an external

The first prototype, YT-27 1300, in its final configuration. (Archive Colonel Aparecido Camazano Alamino)

canopy jettison system, on both sides of the fuselage; redesign and placement of new inspection windows on the wings and fuselage; placing a dorsal fin at the root of the tailfin, to avoid the rudder lock effect; placing fins at the root of the horizontal stabilisers to improve spin characteristics; modification of the ventilation and air conditioning system; changes in the aileron supports; improved cockpit sealing and illumination; introduction of a new circuit breaker panel; installation of engine ignition in the rear cockpit and an increase in the ordnance load from 560kg (1,234lb) to 1,000kg (2,204lb).

'According to the 'Proposal for Preliminary Technical Specification', which is the official contracting document, it provides for compliance with FAR 23. Whenever possible and not mandatorily, other military specifications, among them MIL-S-83691 (Stall/Post-Stall/Spin-Flight Test Demonstrations Requirements for Airplanes). The FAR 23 requirements in this respect are relatively easy to satisfy, more than the military requirements. But the programme that was required by the Comissão de Desenvolvimento, Acompanhamento e Aceitação (CDAA, Development, Accompaniment and Acceptance Committee) may be suitable for a small aerobatic competition airplane, but eventually impossible to execute with a more powerful and heavier aircraft – thus impossible to take place in operation. The demonstration and extensive programme required the 'flat spin' – entry in full power, total reduction after three turns and then coming out of the spin – 'inverted spin' with and without power, among others. These requirements – exaggerated for this type of airplane – have made the T-27 the aircraft that is best rehearsed in spin manoeuvres in its category, with 530 manoeuvres made in total. But just as trials succeeded – and with luck, beyond methodical studies – we might as well have lost the prototype during rehearsals with serious setbacks to the programme. The influence of the French tradition of l'École du Personnel Navigant d'Essais et de Réception (EPNER) in the training of our specialists in the field is excellent and professional, but exaggerated. Since the T-27 and its pilots underwent these tests,

Pre-production EMB-312 c/n 312.005 was used as an Embraer demonstrator for potential clients and for press flights. It later became YT-27 1302 and is currently in storage at the Brazilian Air Force Academy. On the belly can be seen part of the red cover for the air conditioning system, which was removed on the production series. (Embraer)

it is comforting to know, and objective to be publicised, that the airplane is spin-proof', Kovács revealed.

The test campaign also included ground-based vibration – or aero elasticity – when the aircraft was placed on supports, stimulated and analysed by sensors scattered around strategic points of the wings and fuselage. The aircraft was also tested with various configurations of external loads.

One of the most unusual flights was that made without the canopy, with Colonel Cabral in the front cockpit and Sérgio Mauro Costa in the rear. Some resistance tests were conducted on the ground firing four-pound chickens at it, at a speed of 250km/h (135kts). But the team wanted to know how the aircraft would behave in flight without the canopy. Each pilot made a traffic and landing circuit. The trial was a success.

Some weapons trials were carried out with FN Herstal at Zutendaal air base, Belgium. Here, PP-ZTC is armed with two FN Herstal Twin MAG pods, each pod carrying two MAG 58P machine guns with 1,000 rounds per pod, and two seven-tube LAU-32 70mm (2.75in) rocket pods. (Marc Rosenholtz)

After reviewing the preliminary results of the test campaign and monitoring the successful development of the EMB-312 programme, the Ministry of Aeronautics signed a contract for the acquisition of 118 T-27s, plus the option for another 50 airframes, on 14 October 1980. It was a huge source of encouragement for Embraer and for the continuation of the campaign. The FAB would order another 15 airframes from its option, but passed 15 of those on to Argentina so that Embraer could handle the orders. Later, the FAB would request replacement for those 15 from Embraer.

On 23 October 1980, Aviator Day, the YT-27 was taken to the AFA to be displayed to the cadets. The prototype, under the command of Brigadier Menezes, then commander of the Academy, made a 30-minute flight.

According to plan, Embraer performed the inaugural flight of the second prototype, YT-27 1301, on 10 December 1980. The aircraft was manned by test pilot Sérgio Mauro Costa and test engineer Edgard Cavalheiro.

The modifications suggested by Kovács' team, but which had been postponed by a decision of the Technical Board, were made little by little as major events were scheduled. These included demonstrations for clients, for the FAB, or trips for participation in international exhibitions.

'Part of what was left out for nine months from 20 August 1980 to the take-off to Le Bourget in France on 23 May 1981 has been rushed in recent days, with employees working 24 hours a day, forgetting tools inside the prototype and still not being able to adapt the underwing drop tanks. Only one test flight was made the night before the departure for a 20,000km [10,800nm] *ferry'*, explained Kovács. Refinements took until July 1982, naturally causing delays throughout the tooling development process and standard jigs, for example.

By the end of the first quarter of 1982 around 660 flight hours had already been made. But in that year, the programme suffered a first major setback.

On 9 August 1982 the second prototype was lost in a serious accident during a dive to test the design's speed limit. *'The YT-27 1301 was manned by pilot Paulo Thadeu de Oliveira Pimentel and engineer Ricardo K. de Moraes, when it 'broke' when he reached 609km/h* [329kts] *as result of a flutter. The structural limit was later proven in another test flight in early 1983, reaching 608km/h* [328kts], *exceeding the maximum speed for which it was designed. But in the first test, the film that shows the disintegrating prototype was done by an EMB-326GB Xavante that was the chase plane. The crew commanded the rudder to the stop, which resulted in high loads on the tail structure, causing structural failure in the skin near the leading edge. This piece came out and the chain reaction ended up ripping off the wing and canopy. The pilots were almost decapitated by the pressure of the wind on the helmets, which eventually flew off. After that Embraer changed all its helmets to the French standard, which was much safer. With no option but to eject, Pimentel could not reach the handle, because his arm was pressed hard against his chest. He lowered his hand slowly, glued to the body, until he reached the handle with his finger, thus making the ejection. Moraes, who was behind, kept his arm out during the ejection, suffering serious injuries to the shoulder and right arm, with all the ligaments torn'*, said Leber.

On 16 August, seven days after the accident, PP-ZDK took off on its maiden flight. It was, in fact, a pre-production airframe that already incorporated a number of changes and modifications suggested by the in-flight test campaign, as well as structural reinforcements in the rear fuselage section to avoid the problem suffered by the second prototype. This aircraft was used for demonstration flights to customers and also participated actively in the armaments testing campaign. It was already fully integrated with the two underwing drop tanks.

The Tucano completed more than one armament campaign at FN Herstal in Belgium, which were conducted at Zutendaal air base. At least two aircraft went there, PP-ZTC and PP-ZTT. In 1983, during the first campaign, codenamed Liaison, PP-ZTC flew with two Twin MAG pods and two 250kg (551lb) bombs. Each Twin MAG pod was equipped with two MAG 58P machine guns, with 500 rounds each (1,000 rounds per pod). Another test was conducted with four 250kg (551lb) bombs, a configuration that revealed excessive flutter when reaching a speed of 500km/h (270kts) in a 45° dive. The bombs were dropped and the flutter disappeared. In this campaign there was also a test with a Vinten Vicon 70 photo-reconnaissance pod, a system that was not adopted by any of the air forces that operated the Tucano.

PP-ZTT was configured with the Twin MAG pod and two LAU-32 launchers each holding seven 70mm (2.75in) folding-fin unguided rockets.

In Belgium, demonstration and exhibition flights were made at the Pampa test camp in Helchteren. In all campaigns, an armament panel was installed in the cockpit to allow selection and use of weapons by the pilots.

In the early 1990s a new campaign was carried out in Brazil, at São José dos Campos and at the test field of Maxaranguape, in Rio Grande do Norte. These tests were made to introduce and qualify new armaments for the Tucano, meeting the requests of new operators or potential customers who were evaluating the aircraft operationally, such as Colombia and Peru.

With cameras installed under the wings, different types of weapons were jettisoned in flight to ensure they would not collide with the aircraft. There were also live ammunition tests with the heavy machine gun pod, equipped with a 250-round 12.7mm (0.5in) calibre machine gun, and the rocket/machine gun pod with three 70mm (2.75in) unguided rockets. In addition to verifying the operation of these systems, Embraer checked the harmonisation of the sight with these new stores.

With the fuselage still to be painted orange, YT-27 1301 is seen prior to its first trip to Le Bourget, a voyage of 33,000km (17,818nm) that began on 23 May 1981.
(Col Aparecido Camazano Alamino)

The final proof – Tucano on tour

Le Bourget in Paris is home to one of the leading aerospace exhibitions in the world, a place where thousands of people gather to represent operators, suppliers and customers. The Tucano was made for the Força Aérea Brasileira, but Embraer aimed to conquer the international market at all costs.

Paris was clearly an ideal opportunity for the company to demonstrate to the world its revolutionary trainer.

And not only at Le Bourget. For throughout the round trip, Embraer could schedule demonstrations for countries that were interested in knowing more about its trainer.

In addition, it would be an opportunity to test and evaluate the EMB-312 in the most diverse environments, from the hot and humid heat of the Amazon rainforest to the glacial Arctic temperatures.

But making this trip of more than 33,000km (17,818nm) possible would be a huge challenge. Typically, the EMB-312 has a very limited range for long ferries. And, when the voyage began on 23 May 1981, the underwing drop tanks had not yet been adapted. Therefore, it would not be possible to cross the South Atlantic. To get to Paris and participate at Le Bourget, the aircraft would have to take a route via the Americas, Canada, Greenland, Iceland, the United Kingdom and France.

The second prototype, bearing the civil registration PP-ZDZ, was chosen for the mission, and was equipped with an Omega GNS/VLF navigation system. The crew was composed of Colonel Cabral and Commander Brasílico Freire Neto. Accompanying it on the journey was an EMB-121 Xingu, which would be delivered to the French Navy and which carried support personnel and spare parts for maintenance.

On arrival in Belém, in the state of Pará, ice began forming on the trainer's wings when the aircraft was at an altitude of 6,096m (20,000ft) during the descent to land.

Later in the trip, on the leg between Iceland and Scotland, Luiz Carlos Miguez exchanged places with Brasílico.

Arriving in Paris for the 34th Le Bourget International Aeronautics and Space Show, Embraer stood out with four aircraft on display. In addition to the Tucano and Xingu, two EMB-110 reinforced the team.

At Le Bourget many delegations had a chance to get to know the Tucano. Test pilots from some of these nations even flew the aircraft during the event, as was the case with Austria, Egypt, France, Finland, Greece and Tunisia. When a pilot from Sweden

During the second tour, to Farnborough, which began on 10 March 1983, EMB-312 PP-ZTC used underwing fuel tanks: it is seen here at Goose Bay, Canada. The tour included 22 countries, 160 flight hours and 57,000 km (30,777nm). (Brazilian Air Force)

was at the controls, there was a blockage of the elevator control due to a tool that had been left inside the aircraft – a problem already reported by Kovács. But the landing was conducted without problems by Cabral, using a variation of power with flap and landing gear lowered.

While in Paris, the Xingu was delivered to the French Navy and the Bandeirante to PBA – Provincetown-Boston Airlines. As a result, the return to Brazil would only involve the other Bandeirante and the demonstration Tucano.

But before their return, a tragic accident cost the life of Agostinho Moreira Mendonça. The specialist accompanying Embraer was checking the ejection seats' explosive charges when one inadvertently fired. Mendonça received a fatal head injury and the aircraft had to stay in Paris for another week undergoing repairs, while technical teams investigated the episode.

On its return, the Tucano would also be demonstrated in Canada, which had a purchase requirement that could exceed 100 aircraft. But the route between Greenland and Canada was hampered by bad weather, strong winds and there was a possibility that the airfield at Goose Bay would be closed one hour after take-off. On the first part of the trip the meteorology was unfavourable with very strong headwinds, but that later changed to tailwinds, helping the crew. The crews' alternative plan was to eject and wait for rescue in individual lifeboats, floating in the icy waters of the North Atlantic.

Luckily, the crossing was completed successfully.

By the end of its first demonstration tour, the Tucano had travelled 108 hours, coping with snow and cold at temperatures of -25°C (-13°F) and in heat and humidity up to +30°C (86°F).

Farnborough in the UK was the second major international event in which the Tucano participated.

On 10 March 1983 the Tucano started the longest tour the type ever undertook. The fourth flying prototype to be built, PP-ZTC, was accompanied by an EMB-110 Bandeirante, PT-SFA, as support aircraft.

Under the command of Arnaldo Lucarelli Filho and Roberto Werustky, the Tucano visited Venezuela, Panama, Honduras, Haiti, Jamaica and the Dominican Republic on the first stretch that ended in Miami, Florida. Of these countries, Venezuela and Honduras would become Tucano operators.

In the second stage, pilots Guilherme Cará and Sérgio Luiz Brayner Costa Mattos flew to New York, Goose Bay, Reykjavik (Iceland), Prestwick (UK), Paris, Kerkira (Greece), Cairo (Egypt), Khartoum (Sudan) and Kenya.

Now under the command of Arnaldo Lucarelli Filho and Clodoaldo Gualda, the aircraft then flew from Kenya to Uganda, Tanzania, Zambia, Sudan, France, Sweden, Finland, Denmark and then back to France, to participate in the 35th Le Bourget show.

At the end of this epic adventure, 57,000km (30,777nm) had been covered during 160 hours of flight.

The temperature variation was even greater than on the previous tour. The Tucano experienced -45°C (-49°F) in Iceland and +48°C (118.4°F) in Sudan.

A particularly interesting demonstration was flown by Cabral to Libya in 1982, where the regime of Muammar Gaddafi had a requirement to buy up to 200 units.

For this mission, stops were made in the cities of São José dos Campos, Recife, Fernando de Noronha, Ilha do Sal (Cape Verde), Las Palmas (Canary Islands), Seville

(Spain), Ibiza (Spain) Valletta (Malta) and finally Tripoli. For the second time, ice formed on the wings while crossing the Atlantic, at 6,098m (20,000ft). The crew chose to descend to 3,500m (11,400ft) until the ice melted, before climbing again.

In Tripoli, the crew soon became aware of the contrasts of a country ruled by a dictatorship. At the airport, immigration and customs services were performed by soldiers armed with rifles and bayonets. Passports were confiscated and two types of currency circulated in the country. The city was poorly maintained, and the Embraer pilots, despite being in a five-star hotel, could only drive out with escorts.

It is interesting to note that Tucano delivery flights made to Arab countries followed a similar route to that made by Cabral on the trip to Tripoli. The difference was on the stage after Las Palmas, which routed via Alicante (Spain), Kerkira (Greece), Heraklion (Greece), Cairo and then to their final destination in the Middle East. This was a seven-day trip with four rest stops and a little over 40 hours of flight.

The choice of a name and production

The EMB-312 designation given by Embraer to its newest trainer became world-renowned after the aircraft had begun to enter service with diverse air forces. However, while still in the development phase, Ozires Silva believed it was necessary to devise a name for the new Embraer product, to develop an identity for the '312'.

After some discussion among the board of directors, it was suggested that the AFA become involved in the choice of a name through a cultural contest organised by the cadets. Embraer's board of directors then came up with some rules, specifying preference for the name of a Brazilian bird and avoiding words with spelling accents used exclusively in Portuguese.

The first EMB-312 Tucano production line at São José dos Campos (SP), in Hangar F-60. Later the production line was moved to Hangar F-107. (Archive Col Aparecido Camazano Alamino)

With the approval of the AFA commander, the contest was organised and the name was voted. It was Tucano – toucan. The suggestion came from Cadet Carlos Fernando de Souza Panissa, who was in the fourth year, taking an administrative course.

The baptism was made on 23 October 1981, Aviator Day. The ceremony was led by Brigadier Cherubim Rosa Filho, commander of the AFA. Cadet Panissa received a cash prize, a model of the aircraft and received the honour of bathing the nose of the prototype with champagne – the nose already displayed the name Tucano.

Besides money, time was another scarce ingredient at Embraer. As the programme progressed in ground and flight trials, it was necessary to begin producing the jigs and tooling for mass production of the aircraft.

The line would occupy an area of 7,500m^2 (80,729 sq ft) in Hangar F-60. During 1981 and 1982 all the tools were designed, built and the first parts began to leave Production.

'I called the structure staff and together we coordinated with the tooling team to build the airplane as standard. When this airframe was ready, with the 1303 registration, we went to flight preparation and finally delivered it to the FAB', recalled Leber.

The EMB-312 Tucano production line at São José dos Campos, showing Hangar F-107. (Archive Col Aparecido Camazano Alamino)

Table of prototypes, pre-production and demonstrator aircraft

c/n	Variant	Registration	First flight	Fate	Remarks
312.001	A	YT-27 1300	16 August 1980	w/o	First flight prototype; delivered to Brazilian Air Force (FAB) on 8 February 1984; damaged beyond repair after a forced landing near São José dos Campos (SP).
312.002	A	YT-27 1301	10 December 1980	w/o	Crashed on 9 August 1982 with pilot Paulo Thadeu de Oliveira Pimentel and engineer Ricardo K. de Moraes. Aircraft 'broke' as a result of flutter as it reached 609km/h (329kts).
312.003	A			Unknown	Static ground test structure
312.004	A			Unknown	Fatigue ground test structure
312.005	A	PP-ZDK	16 August 1982	Active	Pre-production and demonstration aircraft for Embraer, used at Farnborough in 1982; became YT-27 on 23 February 1984 and delivered to FAB as 1302; remained at São José dos Campos and used for in-flight tests until 2012, when it was delivered to Air Force Academy in July; returned to flight as T-27 in November 2017.
312.007	A	PP-ZTC	14 March 1986	Preserved	Used as demonstration aircraft at Le Bourget in 1983 and Farnborough in 1984; converted into Short Tucano concept demonstrator, made first flight in new configuration on 14 March 1986; was sent to UK with test registration G-14-007, receiving the registration G-BTUC (British Tucano) on 19 June 1986. It was withdrawn from use on 20 December 1996 and moved to Ulster Aviation Society Heritage Centre, Langford Lodge, where it is preserved.
312.053	A	PP-ZTT	18 May 1984	In Angola	Used as Embraer demonstrator. When to Le Bourget in 1987 and widely used for weapons campaign, in-flight tests, demonstration for press and clients. Sold to Angola in 2003 as R-456.
312.149	A	PT-ZTK	Unknown	In Angola	Used as Embraer demonstrator. Participated at Farnborough 1986, 1988, 1992, 1994 and at Le Bourget 1989 (inscription 38) and 1993 (inscription 23). Sold to Angola in 2003 as R-457.
312.161	G	PT-ZTW	28 July 1986	Preserved	Embraer demonstrator for EMB-312G, similar to Short Tucano, to be offered to Brazilian Air Force and other markets as a more capable aircraft with superior combat support. It was modified to EMB-312H, for the JPATS competition in US. The first flight in new configuration was made on 9 September 1991. Now preserved at MAB Museum, at São José dos Campo (SP).
312.454	H	PT-ZTV	15 May 1993	WFU	Although used as a Super Tucano flight prototype, it retained the c/n of the EMB-312 family. Used for flight campaign, withdrawn from use and disassembled.
312.455	H	PT-ZTF	October 1993	Preserved	Second flight prototype of the Super Tucano, also using the EMB-312 c/n. Preserved at Gavião Peixoto (SP) as gate guard.

WITH THE BRAZILIAN AIR FORCE

The Implementation Committee

Two FAB air units would initially be provided with the Tucano: Academia da Força Aérea (AFA, Air Force Academy) and the Esquadrão de Demonstração Aérea (EDA, Air Demonstration Squadron) – popularly known as the Esquadrilha da Fumaça (Smoke Squadron).

Both were of paramount importance to the FAB at that time. The first would use the Tucano for advanced instruction and the other would be the FAB's calling card in Brazil and abroad. The Smoke Squadron would display the quality of Brazilian pilots and industry, as well as inspire aeronautical culture – civil or military – in the young, as all well as advertise the aircraft for sales to other air forces.

But before the first six aircraft arrived at Pirassununga – the headquarters of the AFA and EDA – it was necessary to prepare a whole new structure for their reception.

'It was a first-world process, studied and planned with due time and necessary anticipation, all based on the AFA's vast flight instruction experience. In order to receive the Tucano, the Academy created a commission for implementation of the T-27, made up of experienced instructors who, always working in pairs, would have the task of studying and preparing the data of its characteristics, as well as prepare the text for presentation in technical instruction manuals, the flight manual and even your pocket manual – better known as the checklist,' commened the experienced Colonel Aparecido Camazano Alamino, who was part of the Tucano implementation committee. A captain at the time, he had already flown operationally in the AT-26 Xavante, Neiva L-42 Regente ELO and Bell UH-1H

'Since Embraer was state-owned at the time, all these assignments were assigned to the AFA by the Aeronautics General Staff and by the Research and Development Department, precisely to employ their experience in the subject and to prepare all literature according to academic and educational conveniences of the Academy, since Embraer did not have the experience or personnel to do so,' added Colonel Camazano.

The commission included six instructors and six Smoke Squadron pilots – the latter having priority in flying the aircraft since they were to begin air demonstrations and continue to generate publicity for the new Embraer product.

A few days before the FAB officially received the first six T-27s, the Smoke Squadron pilots went to Embraer in São José dos Campos, to pass through ground school and make five flights each in the Tucano.

'The test pilots showed us everything about the airplane, all its limits. They were not aerobatic pilots, that would be down to us', revealed Major Celso Luiz Cardoso Vilarinho, at the time a pilot with the EDA.

Lieutenant Colonel Geraldo Ribeiro Júnior, commander of the Esquadrilha da Fumaça, was the first to fly solo in FAB 1303, during a 37-minute flight on 8 September 1982. In sequence, other Smoke Squadron pilots flew and qualified in the T-27. Everything was ready for the big day – 29 September 1983.

That date was remarkable for the FAB. At a ceremony that brought together cadets, instructors, officers and other guests at the AFA, six aircraft flew in formation. Four T-27s were painted in red and white, the colours of the Smoke Squadron, and two in white and orange – typical for FAB trainers.

Leading the formation was FAB 1303, which was flown by the commander of the AFA, Brigadier Menezes. The Minister of Aeronautics, Brigadier Délio Jardim de Mattos, accompanied the flight, in the rear seat of FAB 1303.

With the aircraft finally delivered, it was time to introduce the Tucano to the AFA's advanced instruction syllabus. The Smoke Squadron pilots, already qualified, began to train the Academy instructors. In early 1984 the Smoke Squadron dedicated themselves to training and exhibitions, since they were requested to perform in many Brazilian cities and abroad. New officers, trained as instructors in the Tucano, released the Smoke Squadron and assumed the task, concluding implementation work.

'On 14 March 1984 I flew the Tucano for the first time, in T-27 FAB 1312. In that same airplane I was first declared a pilot, on 16 March; and then an instructor, on 17 May,' explained Colonel Camazano.

'Once trained, the pairs of instructors – the most experienced, with flight hours in the AT-26 Xavante and T-37C – were put into action, each in charge of training, writing and rehearsing the manoeuvres and the various flight stages: pre-ground; manoeuvres; aerobatics; instrument flight (basic and advanced phase); formation flight (with two and four aircraft); VFR and IFR navigation; training in all kinds of emergencies, among others. We established the safety parameters that were stand-

The first T-27 delivered to the Brazilian Air Force, FAB 1303 was flown from Embraer to Pirassununga, the Air Force Academy headquarters, by AFA commander, Brigadier Menezes (right). Accompanying him in the rear seat was the Minister of Aeronautics, Brigadier Délio Jardim de Mattos. (Academia da Força Aérea)

ardised for the manoeuvres with the cadets, all being parameters with lower margins than the limits of the airplane.

'Great importance was given to the degree of difficulty that each manoeuvre required, i.e. how much time of flight or preparation the future pilot would spend to perform it correctly and safely.'

In this process it was discovered that some manoeuvres in the Tucano could be learned perfectly with much less training than in the T-37C, for example. In a comparison between the aircraft, the Tucano provided a saving of 20 per cent of the total flight hours at the basic stage. The operator benefited from fuel economy both during training flights and in terms of the total spent for a complete training programme.

'The manuals were always based on the standard of other similar consolidated and tested manuals, such as for the T-6, T-37C, T-25 and even the AT-26 Xavante, having in mind that it was an operational training aircraft with seats arranged in tandem configuration. This process lasted until late 1984 – that is, we can consider a year of preparation, both for the manuals and training new instructors. Over 1,000 hours were flown at this stage. In this period the Implementation Commission created a checklist according to the required logical sequence. The normal airplane check procedures were always performed from left to right, to facilitate understanding for the new pilot. Such procedures even led to the relocation of certain switches, instruments and equipment of the airplane, precisely to facilitate the sequence and the learning of the cadets', added Colonel Camazano.

'Implementation was easy, it didn't give us any problems. That's because we had planning and doctrine. All very well done and unhurried', concluded Colonel José Faria dos Santos, then captain with the Esquadrilha da Fumaça.

Initial problems

Any aircraft, wherever it is built, presents several problems at the beginning of its operational career. Typically these teething troubles are not serious, but sufficient to reduce flying availability or even lead to a temporary grounding of the fleet.

It was no different with the Tucano.

In the early months, among the most frequent problems was the activation of the canopy light, which indicated its possible opening in the middle of a flight. The pilot was then forced to abort training, hold the canopy's operating lever with his hands and land immediately. In most cases it was a false alarm.

'However, one day I was making a loop and right at the top the canopy opened. I was number 3 and ejection was impossible due to the canopy's lateral displacement. I landed the airplane, we removed the canopy and I returned to the AFA on the same day – with a convertible T-27. It was not one of the best flights I've ever done,' recalled Colonel Faria.

'Also we verified a problem of aileron buffeting. The airplane began to flap its wings and this happened on almost every aircraft. An adjustment was made and everything was settled.

'Another problem was that the parking brake lever came out of position as we manoeuvreed using the aircraft pedals, leaving one or both of the landing gear wheels locked. In addition to having a very short and abrupt landing, the tyres exploded. It

happened with me in Farnborough. This problem led us to insert a procedure in the checklist at the time of landing so that it did not occur anymore. With every problem we had to stop all the airplanes, check, test the airplane again and release the fleet to fly again. The airplane was flying a lot and it stopped a lot too. Those were small problems, such as hydraulic and electrical locks, but Embraer kept some engineers in Pirassununga who were accompanying the airplane every day. They repaired everything quickly.'

The Tucano's air-conditioning system initially lacked efficiency.

'We worked hard to get the cooling system running and to do the best we could, but we've always had a lot of trouble with it. It was problematic mainly because the Tucano is not pressurised and has a very large glazed surface – that is, it warms too much or cools too much', revealed Leber.

In fact, the system chosen was the 'freon cycle,' which was heavier, incorporated large and numerous components (condensers, evaporators, and compressor), thus congesting internal space. There was a need for larger air intakes to achieve good efficiency on the ground and at low speed, but the choice fell upon smaller air intakes, without changes to the aircraft's aerodynamics.

'We sent a team to an air conditioning course at Ford, in São Bernardo do Campo, in São Paulo, who on their return made suggestions to Embraer, who remade the project providing us with an improvement', explained Colonel Faria.

Problems of crystallisation of the canopy also occurred, and one almost caused a serious accident. It was during emergency landing training on a dirt track, at the city of Tambaú. The cadet performed a touch-and-go, but did not notice a tall termite mound that was on the track and hit it with the right landing gear. As the mound was very tough, the impact severely damaged the wing, which was pierced by the shock absorber.

'It was decided to spread the fire-retardant foam across almost half the track, as there was not enough foam to spread across its entire length. As the landing would inevitably generate sparks, the aircraft flew until there was minimal fuel left and it could make a perfect landing. The aircraft spun on the broken landing gear and stopped without serious consequences and without touching the foam. The instructor was able to prevent the aircraft from being lost or having to be rescued from the scene of the incident. And it did not take long in the hangar before it was returned to flying condition completely repaired', recalled Sergeant Marcomini.

Structural problems also occurred in the early days.

'The aircraft began to have a problem with the wing fixing bolts, which are integral with the fuselage. The Embraer personnel went to investigate and claimed that we were exceeding the aircraft's limits. But the Tucano is equipped with a fatigue meter, which indicates if the airplane has exceeded its limits. And nothing was recorded in it. But then they warned that the airplane, throughout its operational life, could not exceed -2.5g on more than four occasions. The aircraft's lifespan would withstand -2.5g only four times. And I had done it 170-odd times. It was something that wasn't in the manual and no one had warned us,' said Major Vilarinho.

'The FAB had used the standard of the T-25 Universal instruction flight at the AFA as a basis, thus setting these limits. But the energy pattern of the two airplanes is completely different. With the T-25, for example, I made 10 loops in one hour and with the Tucano I made 40!

'To correct this, reinforcements were applied in the wing, but even so, 10 years later a Tucano wing came off in flight with the Fumaça.'

The star of the show

Only are the Blue Angels of the US Navy are older than the Smoke Squadron.

In the early 1950s, the far-flung Campo dos Afonsos, in Marechal Hermes, west of Rio de Janeiro, was the cradle of Brazilian aviation. It was there that the Airclub of Brazil had appeared at the beginning of the 1910s, the second oldest of the world, only behind the Aéro-Club de France. In 1919 the site became the first home of the Army Military Aviation School. But by the 1950s the FAB was in Campo dos Afonsos, having been established in 1941 with the fusion of Military Aviation, Naval Aviation and the Department of Civil Aviation of the Department of Transportation and Public Works.

Fairchild PT-19 primary and basic training aircraft and T-6 Texan advanced training aircraft took off from its airield. These aircraft and their instructors, from the School of Aeronautics, provided the future military pilots of the FAB.

In February 1952, the School of Aeronautics followed its normal routine. It was lunchtime after a morning of instruction flights with the cadets from the school. All the officers were in the hot and stuffy mess room of Campo dos Afonsos. All but four: Mário Sobrinho Domenech, Haroldo Ribeiro Fraga, Cândido Martins da Rosa and Jayme Selles Collomer. These lieutenants were in the school's hangar, airier and empty at the time. They ate a quick snack, and headed towards four T-6s with orange wings and silver fuselages. Belts were buckled, engines running and propellers spinning. The flight was supposed to take place in secrecy like, all the others they had routinely been conducting. This was not cadet training, not even training in itself. These were precision aerobatics flights – something previously unheard of in the FAB.

Over the then quiet and uncrowded Barra da Tijuca, the pilots began doing basic and solo manoeuvres. Then in pairs, trios and finally in a four-ship formation. Loops, tonneaux and other manoeuvres closely resembled the aerial combats and ballets performed by World War I and World War II aces.

The formation then became tighter, with the aircraft flying closer together.

The inspiration for these flights came from news that had just arrived from the US about the US Navy Blue Angels, who drove crowds to a frenzy with their in-flight manoeuvres.

At the time, this culture was expanding in Brazil. On some special occasions pilots made flyovers or even conducted aerial parades on Independence Day.

For those four youngsters there was great excitement in being able to fly in a squadron set up especially for demonstrations. But everything took place in secret at that time and the flights were carried out between one instruction and another with the cadets of the School. There was no authorisation for this.

But the secret was soon out in the open when rumours about the flights came to the attention of Colonel Délio Jardim de Mattos, then head of the Instruction Division of the School of Aeronautics (and future Minister of Aeronautics). Being an aerobatic enthusiast, Mattos called upon the four pilots and congratulated everyone for their initiative. Not only did he give the green light for the flights to continue, he wanted to see a live performance himself.

The Esquadrilha da Fumaça shows its first livery over the unique Zeppelin hangar at Santa Cruz Air Base. This is the only survivor among three similar hangars in the world. (Archive Col Aparecido Camazano Alamino)

Over their usual flying area, Mattos and then Lieutenant Colonel Aroaldo Azevedo watched the demonstration from the air, in a fifth aircraft. They were very satisfied with what they saw.

Colonel Clóvis Monteiro Travassos, then commander of the School of Aeronautics, was called to attend the presentation and soon approved the creation of the Demonstration Squadron in April 1952.

Four T-6s from the advanced placement collection were chosen by the group and these aircraft were used in the intervals between the aerial instruction.

On 14 May of that year the first official presentation was held over Campo dos Afonsos, a date that is celebrated as the Smoke Squadron's anniversary. Slowly but surely the group began to be requested at civic-military presentations.

Until 1955, the Demonstration Squadron flew with borrowed aircraft finished in the aerial instruction paint scheme – orange wings and aluminium lacquer fuselage. That year, however, given the success achieved with demonstrations that had won admiration for their daring low-pass and formation flights, the squadron received five exclusive aircraft – four for use in presentations and one as a reserve.

With their own aircraft, the now Captain Collomer created a special paint scheme of white, red and blue for the T-6 to stand out in the skies. The first presentation in front of a larger audience and with the new colours occurred on 21 October 1956 above Copacabana beach.

Even with the new paint scheme, it was difficult for those on the ground to follow the T-6s, which at times became black specks in the sky.

The mechanics then adapted a system that injected engine oil into the exhaust of the aircraft. Because of the high temperature, the oil would burn instantly and leave a trail of white smoke in the sky, demonstrating the aircraft's position to those on the ground. It was a solution created entirely in Brazil, without help from the manufacturer. Soon the team received the affectionate nickname Esquadrilha da Fumaça – Smoke Squadron – a name that was then officially adopted.

At the same time a fifth aircraft was added to the demonstration flight to entertain the audience while the other four reorganised after each manoeuvre and prepared for the next pass, giving a sequence to the demonstrations.

With the increase in popularity and demand for more presentations in Brazil, Order No. 1,049/GM3 dated 22 October 1963 officially created the Smoke Squadron as an independent air unit, separating it from the School of Aeronautics and establishing its headquarters at Santos-Dumont Airport, in Rio de Janeiro.

As the team reorganised and became independent, other changes soon became possible. In October 1964 a first air demonstration with eight aircraft took place and four years later the veteran T-6s gave way to the more advanced Fouga CM.170-2 Super Magister training jets. Seven were acquired new from the factory in France, receiving the colours of the national flag on their wings and fuselages. On 18 April 1969, the first exhibition was held with the new jets, which were more manoeuvrable and powerful. But despite being more modern and having these advantages, the aircraft did not adapt well to the conditions found in Brazil. In addition to their limited autonomy, the jets could not land on most of the country's dirt and grass tracks (only on asphalt and concrete) from which the T-6 could operate easily. In addition, their engines required the use of kerosene instead of aviation gasoline. This was a complication because not all localities had kerosene easily available at the time.

The solution was to return to the T-6 for distant cities with little infrastructure and leave the Magister for displays in larger urban centres. In 1973 and 1974 the jets used coloured smoke in the colours of the national flag, but in 1975 they were deactivated and sold back to France, due to their lack of flexibility for operations in locations with limited structure.

The following year the Smoke Squadron hit another hurdle. The Aeronautics Staff elected to deactivate the entire T-6 fleet due to its obsolescence and the difficulty in finding replacement parts for the aircraft. The solution was found in the EMB-326GB Xavante advanced training and attack jet, an Italian design now being built under licence in Brazil to equip several FAB units.

The Smoke Squadron would be a great opportunity to demonstrate the quality and evolution of Embraer, a company founded in August 1969 and already engaged in projects as complex as this. Four aircraft were assigned to Fumaça and the team began

Perhaps the most attractive colour scheme of the Esquadrilha da Fumaça was this red one. This is the team's number 6. (Luciano Porto, www.spotter.com.br)

to receive flight instruction at the 3rd Mixed Reconnaissance and Attack Squadron at Santa Cruz Air Base in Rio de Janeiro. But because there were still few Xavantes available and due to budget constraints, it was decided to temporarily deactivate the Smoke Squadron.

Their final presentation took place on 31 January 1976 in the city of Piquete, in São Paulo state. In all, 1,225 displays were made with the T-6 and 46 with the Magister.

Order No. 567/GM3, of 8 June 1977, deactivated the Smoke Squadron. Their return would happen only when the economic situation allowed it and preferably if the aircraft used could be of Brazilian manufacture.

This break lasted seven years. During this period other short-lived demonstration teams emerged, albeit without the Smoke Squadron's reach. These included the Alouette Squadron formed with the Xavante – curiously the same aircraft that was to equip the Smoke Squadron but was abandoned for economic and availability reasons – of the 2°/5° GAV 'Esquadrão Joker' (2nd/5th Aviation Group 'Joker Squadron'), from the Centro de Aplicações Táticas e de Recompletamento de Equipagens (CATRE, Tactical Applications and Crew Readaptation Centre, 1 January 1974 until 1989). Based in Natal, this unit was responsible for training FAB fighter pilots.

In addition to this was the Daedalus Squadron, 1°/5° GAV 'Esquadrão Rumba' (1st/5th Aviation Group 'Rumba Squadron') that was also part of CATRE. Using Brazilian-made twin-engined Embraer C-95 Bandeirante turboprops, the squadron trained military pilots for transport and patrol missions. These aircraft were also fitted with smoke generators and performed in many Brazilian cities.

There was also the Joker Squadron, which lasted longer and actually appeared when the Smoke Squadron was still operating. Created at the AFA, its first presentation took place on 10 July 1969 using nine T-37C advanced trainers (two pilots per aircraft). These aircraft were used to train the cadets at the AFA. None of the jets received a special paint scheme or insignia and their deactivation occurred in 1977 with the retirement of the T-37C from active service. In all, 200 presentations were made, including one in Paraguay.

The return of the Smoke Squadron was heralded in 1979, when a group of pilots informally began performing aerobatic formation flights over the Academy.

In May of the following year, Brigadier Menezes, now at AFA command, asked the Training Division about the possibility of creating a new Academy squadron using the T-25 Universal.

The then commander of the 1° Esquadrão de Instrução Aérea (1° EIA, 1st Instruction Air Squadron), Major Toscano, requested Captain Dailson, the Operations' Chief, to form the squadron using the most experienced instructors.

Lieutenant Brito was the flight leader with Captain Dailson on the right wing, Lieutenant Vidal on the left wing, and Lieutenant Nazario as the slot, while Lieutenant Vilarinho was the solo.

The squadron had five aircraft and the first presentation took place on 9 July 1980 in the presence of Brigadier Menezes. The next day the first official demonstration took place, at the ceremony marking the delivery of the first year AFA Cadets' rapiers.

Informally the team was known as the AFA Squadron and a few presentations were made that year. The last one took place on 10 December 1980, at the graduation party of the fourth year cadets, under the command of then Lieutenant Faria and also including Lieutenant Bragança, Captain Dailson, Lieutenant Sérgio Sales and Lieutenant Vilarinho.

The full formation during a training session over the Brazilian Air Force Academy. (Giuseppe Ricardo Passarini)

This was the end of the AFA Squadron, but the arrival of a new member became, in a sense, one of the catalysts for the creation of the future Smoke Squadron. This was the then Major Geraldo Ribeiro Júnior, who had flown with the original Smoke Squadron in the T-6.

'I taught Junior Ribeiro to pilot the T 25. And then we showed our presentation to him. And he called us all turds: he said we were flying too far from each other and that was wrong. Then he gathered everyone together. Very close, really damn close. It was even scary', recalled Colonel Faria.

With the introduction of Major Ribeiro Júnior, Esquadrão Cometa Branco (White Comet Squadron) was created. Comet was the radio callsign of the 1° EIA. White referred not to the colour of the T-25, but signified that the name was left blank – it had not yet been defined. The aircraft did not receive a special paint scheme or markings.

The culture, the safety, the types of high-precision manoeuvres and other Smoke Squadron characteristics were introduced in this squadron by Ribeiro Júnior.

On 31 March 1981, White Comet's first presentation was held at the AFA to mark 17 years of the 1964 Revolution in Brazil, which deposed President-elect João Goulart, who threatened to establish a Communist dictatorship in the country.

On 22 April 1982 the aircraft began to use smoke in the presentations, a system created by the squadron's mechanics, who were now responsible for six aircraft (previously five).

White Comet's situation began to change with the accident suffered by the second Tucano prototype, YT-27 1301, on 9 August 1982. To reaffirm national industry, Minister

Délio Jardim de Mattos decided to recreate the Smoke Squadron using just the Tucano. In addition to showcasing Embraer and its capabilities, it would promote the aircraft commercially and globally through air demonstrations. It was exactly what was foreseen in the 1977 order that deactivated the Smoke Squadron – the use of an aircraft of Brazilian manufacture.

The re-establishment took place on 21 October 1982 by Decree No. 87,739, using the pilots and mechanics of White Comet – this team would make its last appearance on 7 August 1983. Other officers and graduates were also called.

White Comet performed 55 demonstrations in total, from March 1981 to August 1983.

'We had the whole routine of flights, preparation and course in the T-27. But in parallel we had to choose and decide other things, like painting the airplane. How are you going to paint the airplane, what colour? We entered a room with cardboard on two walls with paint suggestions. While I was doing the theoretical course I spoke with the artists. They were a Belgian and a Brazilian (Anne and Coimbra) and we were talking and arranging the details', explained Colonel Faria.

There were more than 200 proposals, and the 20 most interesting were chosen and applied on scale models for better visualisation. The choice was made by Brigadier José Pessoa Cavalcante de Albuquerque, commander of the AFA, who admired the Northrop F-20G Tigershark prototype's red paint scheme.

After receiving its aircaft in September 1983, the Smoke Squadron began intense preparations for its forthcoming presentations. Then Lieutenant Vilarinho was a lover of aerobatics and 'taught' the FAB to perform inverted flying, as well as the 'Lomcovak' manoeuvre (which includes a climb followed by three turns of the aircraft on the wing's own axis, a descent and recovery). Vilarinho was one of the key elements within the group and these manoeuvres were part of the White Comet's demonstration.

'On one occasion, I remember asking to fill my airplane with enough fuel for two hours of flying, because I wanted to explore many manoeuvres at that time. I was very curious and even wanted to test the Tucano's performance. But after 35 minutes of flying I landed and people thought the airplane had problems. In fact, it was me who had problems; I could not stand the Tucano's energy because you get exposed to G loads for more time. In the Tucano I've flown inverted for 13 minutes before the engine stopped. In the T-25s' case, five seconds inverted and the engine went off. I made inverted loops and other situations that I had not done before', explained Major Vilarinho.

In just over two months the demonstration was ready and the Smoke Squadron was first presented on 8 December 1983 at the AFA during the graduation of the fourth year cadets.

On 27 February 1984 the Smoke Squadron began the trip to participate in its first international exhibition with the Tucano. It would take place at FIDA (Feria Internacional del Aire, Air International Fair) in Santiago, Chile. The month before the aircraft received two smoke generators installed under the wings, each one weighing 57kg (126lb). The system was efficient but caused high aerodynamic drag and had a negative effect on manoeuvres.

From Pirassununga, supported by C-95A FAB 2296, the six T-27s went to Santa Maria and the next day to Uruguaiana in Rio Grande do Sul and Mendoza in Argentina.

The revised paint scheme in the colours of the national flag. (Sgt Johnson Barros, FAB)

On 1 March they crossed the Andes mountains and over the next two days undertook two training sessions in Santiago before the official presentation with five aircraft on 3 March.

At the beginning there were no differences between AFA and Smoke Squadron aircraft. But then the mechanics of the EDA adapted a 40-litre (10.5-gal) oil tank in the luggage compartment and a hydraulic pipeline that terminated in the Tucano's left exhaust – following the same principle used in the T-6. It was lighter and did not produce the drag of the old external smoke generators. It improved the aircraft's performance and allowed the introduction of new manoeuvres. A first presentation with the new system was made on 9 August 1985 in Abbotsford, Canada.

Another milestone in the Smoke Squadron's history occurred in 1984. This was the team's first intercontinental presentation, in the skies over Europe, including a daring crossing of the South Atlantic. The trip would involve two presentations in Portugal and one in Germany.

A number of steps were needed in order to prepare the Smoke Squadron for this long-distance ferry, including sourcing survival radios, maps, navigation charts, information on Europe, hotel reservations, transport, installation of specific navigation systems, passports, publicity and PR materials, spare parts and other items.

A break performed over the South Atlantic Ocean. (Esquadrilha da Fumaça)

Lockheed SC-130E Hercules FAB 2459 of the 1°/6° GAV 'Carcará Squadron' (1st/6th Aviation Group 'Carcará Squadron') would follow and support the squadron on the crossing and during presentations in Europe.

Unlike the trip to Chile, ferrying across the Atlantic would require the use of drop tanks.

'We did not know if the underwing fuel tanks would properly transfer fuel to the wings. Ribeiro Junior did not like doing it, but I took the airplanes, all seven that we had, and prepared a huge flight plan that no one understood at the end. The mission was to understand if the fuel would be transferred correctly. Embraer did not know. I was the flight leader and we made a circuit that had a six-hour flight. The order was: whoever had a problem with fuel transfer should speed to São José dos Campos, so that Embraer would fix it. The seven airplanes landed there', recalled Colonel Faria.

Four aircraft had trouble transferring fuel. Two showed flaws in the Omega navigation system.

The Tucano's drop tank was adapted from that of the Xavante, which was pressurised. In the Tucano, fuel transfer was made using a fuel pump. However, the electrical connection unplugged during flight.

'They fixed it and I made another flight. This time alone and manoeuvring the aircraft in the G limit, -2 and +4, and then it showed no problems', added Faria.

The Smoke Squadron in a 'snake show' formation over the Amazon rainforest. (SO Waldemar Prieto Júnio, Esquadrilha da Fumaça)

Take-off from Pirassununga to Europe was at 09.15 on 11 August 1984, arriving in Recife, Pernambuco, 6h 10min later, when the pilots had already tested the drop tanks, navigation, communication, oxygen, engine and hydraulics. Besides the seven T-27s of Smoke Squadron, Tucano PP-ZTT from Embraer joined in Recife, since it would participate in demonstrations in Europe. This aircraft contributed to the demonstration and evaluation campaign for the competition that would supply the UK with the Tucano.

The next day saw a 1h 40min flight to the archipelago of Fernando de Noronha. Around 22.00 the participants in the mission, including the Smoke Squadron, SC-130E and Embraer Tucano crews, received a weather briefing on the conditions for next day's crossing.

Everyone got up at 02.30 on 13 August with take-off at 04.00 due to the Intertropical Convergence Zone, a cold front that exists in the passage of the Southern to the Northern Hemisphere and that causes severe turbulence. This zone is constant and the sooner take-off happens, the better.

'We had no GPS and no radar. It was a 6h 55min flight without standing up and walking, sitting in the ejection seat. The Tucano is equipped with a relief system for the pilot, a funnel connected to a hose. But it was not comfortable because when you start peeing the flow returns. The ideal procedure is to pee a bit, stop, wait for the liquid go down and then resume. But I ask: can you do this? Of course not! The solution was a 2.5-litre [84 fl oz] *Coke bottle that fits any size and holds a good amount. Someone once took a dump – in a lunch box!'* said Colonel Faria.

The Tucano had eight hours of fuel autonomy, but five hours of oxygen. From Fernando de Noronha to Ilha do Sal, one of the Cape Verde islands, it would be a flight of seven to eight hours, depending on the wind.

The flight was made at 3,900m (12,795ft) so the crew would not use oxygen for 2h 30min. Cruising speed was around 352km/h (190kts). With the aircraft lighter, the formation ascended to 5,500m (18,000ft) and then to 6,700m (22,000ft) and the pilots then used the oxygen masks.

If Ilha do Sal was closed, the viable alternative was Praia, capital of Cape Verde. However, due to the proximity of the two airfields, the chances of both being closed were very high.

Instead of ejecting over the sea, the pilots had the alternative to climb to nearly 9,144m (30,000ft) and, when they were 30 minutes from Dakar, cut the engine and electrical system, retaining only the artificial horizon. At 914m (3,000ft) the engine would be restarted for the approach to Dakar.

'The last three days before the trip were crazy in order to assemble everything, since the C-130 carried propellers, the Smoke staff, spare parts and also a multitude of items. It was a squadron displacement as if we were going to a war. I assembled the PR kits with gifts, bottles of Brazilian cachaça to be given as a present, brochures, posters and lots of things,' recalled Vilarinho.

'I was in great stress, had slept little and was very tired. The first stop was in Recife, after six hours of flight. On the next day, a 1h 40min flight to Fernando de Noronha. It was a very demanding day and I went to sleep at 23.00 and woke at 02.30 in the morning.

'The result was that during the flight to Sal I started to doze until I was fully asleep. I switched off completely.

'I woke up with the team shouting in the radio, calling my name and my nickname, which in the squadron was Baleia [whale]. *I finally woke up and was too low and 90° off the route where I should be. The squadron was out of my reach. I had no navigation system, was alone and lost in the middle of the Atlantic. I entered the same frequency of the C-130 radio and managed to locate the course that I should follow, to find everyone again.'*

On 14 August the flight from Sal to the Canary Islands was another 3h 45min and 15 August was dedicated to maintenance.

From the Canaries, the team went to Base Aérea No. 1, at Sintra, headquarters of the Portuguese Air Force Academy, on 16 August, a 4h 5min flight.

During the first training for the presentation, on the 17th, the front landing gear door of Vilarinho's Tucano opened in flight, and was noticed by all who were on the ground.

For Vilarinho, who manned T-27 FAB 1311, the yellow alert light for lack of hydraulic pressure went on and he lowered the landing gear as an emergency.

The nose landing gear lowered but did not lock.

'I made several manoeuvres trying to lock the gear, but there was no way to make it work. José Faria tried to touch the nose gear with his wingtip to lock it, but Ribeiro Júnior was crazy about the idea and ordered him to stop.

'I declared an emergency, firefighters positioned themselves and spread foam on the runway. I cut the engine, feathered the propeller blades and it spun slowly. A blade barely missed the ground but the next dug itself in and held the nose up with-

out a scratch on the engine cowling. Damage was minimal, but we had a reserve propeller. During the night the mechanics made the necessary repairs and on the next day the airplane was tested by me'.

On the 20th the first official presentation was made in Portugal, at Sintra.

The next stop was in Paris, where at 15.00 on 22 August the aircraft arrived and remained in France for five days. A presentation in Paris had been expected, but it was prevented by bad weather. On the 27th they went to Cologne-Bonn International Airport, in West Germany, to attend celebrations marking 75 years of the airport and 25 years of the local flying club. The next day they began the return to Brazil.

The professionalism and expertise of the pilots of the Smoke Squadron were not only proven in air presentations, but also in emergencies.

'Boss, I have a fire on board. I will eject.'

'Seven, confirm.'

'OK. Confirming fire.'

'Cut engine.'

'Engine off. Fire continues, I'll walk away.'

This dialogue caught between Colonel Faria, the formation's solo (number 7) and then Lieutenant Colonel Américo Soares Filho, the leader of Smoke Squadron (number 1), sounds more like a casual conversation. But it was the most extreme situation Faria experienced in his whole life.

'I was flying as number 7 and started the Lomcovak manoeuvre. In the first turn the entire propeller came loose from the shaft – and I mean entirely. It all came off. And the fire alert light lit on the panel. This is the only light I cannot push a button and turn off, the only one. All others I can turn off. The burning oil leaked with pressure and started melting the canopy. It was total chaos. When the fire alert light is on the only option is to eject. But how to eject and let more than two tons of airplane crash in flames, with chances of hitting the public? That was not an option.

'There were people on both sides of the runway. The airplane was still controllable, but it was just a glider.

'I began to use the airplane's pedals to see sideways and thus lead Tucano where I wanted it to fall. And I did it.

'I ejected at the 'half eucalyptus height', around 23m [75ft], *and heard several bangs beneath me as soon as I pulled the ejection handle. Impact on exit is +20g but I did not lose my helmet, something that happens often in these cases. And I hit the ground hard and fast, because I was too low.*

'I spent two months in a cast, immobilised. But I did not have to read newspaper headlines saying a Smoke Squadron airplane crashed leaving many dead. And I could sleep at night. In a way, the plaster was cheaper', he recalled.

The accident's cause was linked to the fact that the propeller shaft of the Tucano – an aerobatic aircraft – was the same used in the Bandeirante, which of course does not endure the same forces during flight.

The propeller shaft of all EMB-312s had to be modified and replaced with a new one. The Companhia Eletromecânica de Petrópolis (CELMA, Electromechanical Company of Petrópolis) performed all repairs and replacements, while the AFA was responsible for removal and installation.

Another accident, this time at Santos in São Paulo, at a time when the Smoke Squadron was led by Faria, revealed another structural problem.

'In Santos I heard Barreto's desperate voice. It was 16 November 1996. In that area there are many buzzards and I thought he had struck one. But he was the solo, the number 7. We were about 5km [2.7nm] *away and I told him to cut off the smoke because I could not see him well. He repeated that he was not going to give up and then I ordered him to eject. I throttled up to 100 per cent and went as quickly as possible, exceeding the speed limit, since I was higher than him and by diving I was exceeding the limits. The lights on the panel and sounds in the cockpit indicated that constantly.*

'I saw the bobbing tail on the water and a lot of people on the beach surrounding a person. He ejected successfully, thankfully.

'We had nothing on us. No money, documents, wallet, toothbrush... Nothing! It was business as usual: to leave Pirassununga, do our show and go back. But with that accident the Tucano fleet stopped and the airplanes stayed wherever they were', recalled Faria.

The Tucano lost its right wing, which was completely torn from the fuselage. The rest of the squadron landed at Santos Air Base and waited for an Embraer team to inspect each one of the aircraft.

'The joint which connects both wings in the centre broke and with that the wing came off in flight. Using ultrasound and without disassembling the aircraft we checked every one of them – and two showed chances of having cracks. These were taken to Embraer, which made the replacement parts. In other FAB airplanes the cracks were found and repaired. Embraer then decided to design a new mount and send teams to install them on all aircraft operating around the world', said Leber.

Two years later the EDA made history after breaking a record while flying 10 aircraft inverted for 30 seconds at 914m (3,000ft), on 23 October 1996. That feat entered the Guinness Book of Records in May 1999. It was repeated on 18 May 2002 but with 11 aircraft, to commemorate the 50th anniversary of the team's creation. And on 29 October 2006, the centennial of Alberto Santos-Dumont's 14-bis flight, the Smoke Squadron flew 12 aircraft inverted for 30 seconds.

The team remained very active until 2000, when the flying stopped and the aircraft went through an extensive technical overhaul to reinforce the wings that already showed a high degree of wear due to the number of flight hours. The new wings followed the same pattern as those on the Tucano sold to France, which had a full wing spar, among other reinforcements.

On 25 February 2002 the team returned to active service in its first appearance sporting a new paint scheme which comprised the blue, yellow and green of the Brazilian flag – a more patriotic finish. Also new was an automated skywriting system.

At that time the team had crossed the Atlantic twice for demonstrations in Europe, including Germany, France, Portugal and the UK. On the American continent, the Smoke Squadron had visited Argentina, Bolivia, Canada, Chile, Colombia, Costa Rica, Dominican Republic, Ecuador, Guatemala, Guyana, Honduras, Paraguay, Peru, Suriname, Uruguay, the US and Venezuela.

On 31 March 2013 the Smoke Squadron performed a last display with the EMB-312 Tucano, after almost 30 years with the squadron and 2,363 presentations. The demonstration team began another recess period to begin the transition to another new aircraft – the EMB-314 Super Tucano.

At the Academy

At the Academia da Força Aérea, the Tucano was delivered to the 1º Esquadrão de Instrução Aérea (1º EIA, Instruction Air Squadron), the 'Comet Squadron', which had four flights: Antares (black), Vega (green), Sirius (blue) and Castor (red). The aircraft were not divided by flight and no aircraft was exclusive to them.

The cadets begin to fly the Tucano in the fourth year of instruction, with an average of 125 hours per student. As soon as the student reaches the Tucano in their last year at the Academy they will learn more advanced techniques in a faster aircraft with ejection seats and with the instructor seated not to their side, but in the rear seat. These characteristics combine to make the advanced flying more difficult.

Divided into seven phases, the student will convert on to the Tucano and start on pre-solo, where they will learn to take off, conduct the normal and emergency procedures, basic manoeuvres and landing. When the instructor judges their student to be ready, the solo flight will take place.

In the second and third phases, the student begins manoeuvres and then aerobatics. Again, during these phases, the student also will fly solo when the instructor allows. After that, during the fourth phase, comprising IFR flights, the student enters the world of on-board instruments to take off, complete some basic circuits and then land.

The fifth phase is formation, and the student will fly in two- and four-ship formations. Once again, the student will fly solo, but never as a leader. This is the most difficult phase of the course.

The sixth phase is night flights and the last phase is the navigation, in which the student will fly long distances. First they will learn how to navigate using references on the ground like rivers, cities, mountains, etc., both at high altitude and low level (200m, 656ft). Then, the student will navigate using only on-board instruments, and will complete some descent procedures into airfields.

The EMB-312 simulator at the Air Force Academy. (João Paulo Moralez)

A Brazilian AFA instructor prepares for another training mission. (João Paulo Moralez)

During the advanced phase, training flights include three- and four-ship formations. (Sgt Johnson Barros, FAB)

In the 1980s, before reaching the Tucano, the cadets had a first year dedicated to theory, then a second year of basic instruction in the T-23 Uirapurú of the 2° Esquadrão de Instrução Aérea (2° EIA, Instruction Air Squadron), the 'Apolo Squadron' with its four flights. A third year was followed by another theoretical period and the fourth year was dedicated exclusively to flying, in the T-25 Universal of 1° EIA.

However, from 1985 the structure would be quite different. The T-23 was finally retired from the FAB, and the T-25 went to the 2° EIA for primary and basic instruction of cadets in the second year, and the T-27 Tucano took on advanced instruction at 1° EIA for cadets of the fourth year.

To make this transition and move on without delay in their first year of advanced instruction, on 26 September 1984, six students of the third year flew solo in the T-27.

In 1983 a batch of 12 T-27s were already delivered, followed by another 12 in 1984. By the end of 1985 there were 62 aircraft in operation at the AFA. In 1986 the AFA reached the highest T-27 allocation in its history – 67 aircraft. Earlier, the average fleet availability was 70 per cent, rising to 85 per cent according to the maturity of the operation. The AFA flew an average of around 25,000 hours per year of instruction with the Tucano. On 19 September 1986, the Tucano reached 1,000 hours of flight at the AFA, then under the command of 1st Lieutenant Maurício José Gusman Filho.

On 5 April 1987 the Tucano suffered its first fatal accident, claiming the life of 1st Lieutenant Mauro Luiz Ribeiro Pinto, who impacted the ground on board FAB 1343.

A curious fact is that the AFA also had its own mixed demonstration flight squadron. This was the Hybrid Comet, formed by a T-25 and a T-27. The aircraft were flown by AFA instructors who performed inverted flight on the wing, lomcovak and other manoeuvres. All were part of the FAB flying cadet's curriculum.

The squadron was born in April 1993, when the AFA was invited to participate in the anniversary of the city of Catanduva, in São Paulo, on 17 April 1993.

'*Air Brigadier Fernando Mendes Nogueira authorised an entourage, in which he was present. We went on two teams: one C-95 Bandeirante support team and other with the T-25 and T-27 aircraft, whose purpose was to enhance the event with a small demonstration, to be made separately for each of the aircraft. In the afternoon, in the middle of the air events sequence, there was a take-off of each AFA instruction aircraft for the aforementioned demonstration. Brigadier Nogueira had remained on the ground talking with friends of the flight club. The T-25 (FAB 1921) was piloted by Major Willie R. Monteiro de Carvalho and the T-27 (FAB 1335) by Major José Roberto Borin. Showtime comes and, instead of two separate presentations, the T-25 was positioned as leader (with the engine at full power all the time) and the T-27 entered the wing for the flight premiere of 'Hybrid Element"*, recalled Major Willie.

The duo continued flying until 7 April 1995, with 53 presentations in total including four in Brasilia.

'*The T-27 gave more to FAB than we expected. It is a great trainer. The time we received the Tucano was extremely appropriate; it exactly filled the gap that was missing. By the time the aircraft entered service, there was no doubt it represented the training airplane for the next 35 years,*' added Brigadier Menezes.

The AFA also has two T-27 flight simulators at its facilities, which are used in the pre-flight phase, when the cadet has their first practical contact with the aircraft. Normal and emergency procedures are practised in the two-axis simulator.

The simulator was developed in the 1980s with national hardware and software, using few imported components. It took around 30 months for its development, involv-

An EMB-312 flight line, prepared for night flying (João Paulo Moralez)

ing Embraer, the Research and Development Institute of CTA and the private company ABC Simulators and Avionics.

Each simulator had an estimated cost of USD 1 million, and both (AFA-I and AFA-II) were inaugurated in December 1991. The first instruction in these systems took place on 6 March 1992.

On 18 April 2000 the AFA received the first two T-27s with a new paint scheme – similar to that used in France. The new scheme made visual reference easier during formation flying.

By September 2017 more than 553,526 hours had been flown and the Tucano allocation at the Academy comprised 36 aircraft, with 25 on the flight line and the remaining 11 in routine maintenance.

Aerospace Technical Centre

Concurrent to introduction of the EMB-312 with the AFA and the Smoke Squadron, the Aerospace Technical Centre also received and started operating Tucanos through its (Pesquisa e Ensaios em Voo (PEV, In-Flight Research and Testing) unit. This is in charge of research, development, testing and homologation in the area of aerospace, and the unit uses its own aircraft and highly qualified test pilots.

When the Tucano was first delivered to the FAB in 1983, the PEV received prototype YT-27 FAB 1302, previously PP-ZDK, on 27 February 1984. Subsequently, also in 1984, production airframes FAB 1314 and 1321 also joined of its fleet, and remained there until 11 April 2012.

PEV contributed extensively to the Tucano development programme with the initial campaigns and especially during tests of a variety of weapons.

At the end of 1984, using the Marambaia firing range in Rio de Janeiro, YT-27 FAB 1302 dropped 112 Mk 76 bombs (11kg, 24lb, with smoke cartridge), 162 SBAT 37 rockets (37mm, 1.45in, fired from the LM 37/7 launcher) and 162 SBAT 70 rockets (70mm, 2.75in, LM 70/7 launcher). In March 1985 around 3,900 rounds were fired from two C-2 type pods, each equipped with an FN MAG 58P 7.62mm (0.3in) calibre machine gun.

Although using the designation T-27, this YT-27 FAB 1302 was widely used for the training of test pilots, as a weapon test platform and for certification during the 1980s. (Marco Aurélio do Couto Ramos, www.spotter.com.br)

The results were used to produce the armament manual, which was prepared by PEV, and the unit also performed external load separation tests with the YT-27. An unprecedented development in Brazil was the installation of a pod in one of the wings equipped with two high-speed cameras. The system was developed by PEV and allowed the images to be captured by the aircraft itself, without the use of a chase plane.

A Brazilian 45kg (100lb) incendiary bomb on the outer pylon for test purposes. Its tail shows some sign of damage. (IPEV)

On 20 October 1986 the Tucano demonstrated its full capability through a live-fire demonstration at the firing range near the city of Silvânia, in the interior of the state of Goiás. The Yellow Test Squad was led by YT-27 FAB 1302 that launched 28 SBAT 70 rockets using four LM 70/7 launchers; FAB 1385, which fired the same number of rockets; rockets; the third T-27 dropped four BAFG-120 (120kg, 264lb, similar to the Mk 81), bombs and T-27 FAB 1386 dropped four BAFG-230 (230kg, 507lb, similar to the Mk 82) bombs. The latter was the first time a Tucano had carried and dropped bombs of this size.

On 13 November 1990, YT-27 FAB 1302 and T-27 FAB 1321 were manned by Brazilian and Venezuelan pilots, since the latter country was evaluating the purchase of rockets to arm its aircraft. Using the Marambaia firing range, the aircraft conducted several sorties launching the SBAT 37 and the final evaluation of the Venezuelan military was very positive.

'*The Tucano was widely used to do initial training, in-flight test drills, and operational maintenance flights. The aircraft is an excellent training platform that is very easy to pilot and is used for the development of the test course for stall and spin exercises, including inverted spin*', recalled Colonel Luiz Alberto C. Munaretto, who was a test pilot with the FAB.

The T-27 was also used in propeller aircraft performance instruction, demonstrations of test techniques and to assist exercises by students who prepared in-flight test reports using data collected on the flight with the Tucano.

'*It was also widely used on piloting quality flights as a weapons platform – but no weapons were used. One of the most exciting and interesting phases played in the T-27 was the spin and the screw chase. Most were missions with approximately eight spins of six laps with training of diverse entry, escalation and exit techniques from this manoeuvre. All were accompanied by another T-27, who watched the*

Seen here in 2006, YT-27 FAB 1302 was still wearing the old colour scheme. (João Paulo Moralez)

behaviour and controlled the safety. Both types of flight were very full of adrenaline, but also full of lessons for the test pilots in training. The Tucano was an excellent aircraft with characteristics that enhanced crew instruction (pilot and engineer) of flight tests', commented Roberto Follador, another FAB test pilot.

Lagoa Santa Aeronautical Materiel Park

All FAB units are able to perform most of the simplest maintenance of their aircraft directly on the flight line, while tasks that are somewhat more involved are undertaken in the squadron hangars. Meanwhile, the Aeronautical Material Parks are centres responsible for major inspections such as the replacement of large components, extensive structural overhaul, complete painting and other more complex services.

The FAB has a number of these centres across the country, sharing the overhaul and routine maintenance of all models in operation. For the EMB-312 Tucano, these services were assigned to the Parque de Material Aeronáutico de Lagoa Santa (PAMA-LS, Lagoa Santa Aeronautical Materiel Park), located in the region of Belo Horizonte, in the state of Minas Gerais.

Through Reserved Bulletin No. 15, dated 13 September 1982, issued by the Diretoria de Material de Aeronáutica (Aeronautical Materials Board), PAMA-LS concentrated the technical materials, manuals, tooling, parts inventory and other fundamental components and systems to maintain the operation of the Tucano within the FAB.

The first EMB-312 that passed through PAMA-LS was FAB 1319, on 4 June 1984, which had suffered damage on the leading edge of one of its wings after colliding in flight with a bird.

After suffering extensive fuselage damage, T-27 FAB 1332 was recovered by a FAB team at PAMA-LS in November 1990. Here, work is performed around the engine and cockpit panels.
(PAMA-LS)

PAMA-LS also retrofitted 42 FAB Tucano airframes. After the EMB-312 entered service with the Air Force Academy and the Smoke Squadron, Embraer detected some aspects of the design that would need to receive some kind of modification.

'The aircraft continued evolving in the in-flight test campaigns while Embraer was delivering them. In addition, there was the FAB's own feedback about problems that needed correction. We then set up a group to evaluate each airplane that had already been delivered to the FAB and to introduce the necessary changes therein. This was extended to almost all of the FAB's fleet, but the number of modifications was decreasing in the newer airframes, since many changes were introduced while still in the production line. Since the first were from the Smoke Squadron, and it was a priority, we started with their airplanes, which had been delivered first. The airplanes arrived in the flight preparation hangar in São José dos Campos where we set up a production line with six stations. The second step was to bring FAB personnel, sergeants and non-commissioned officers to learn on the spot, and the third step was to go to Lagoa Santa Aeronautical Material Park to introduce the line of modifications there', explained Ary Leber.

A total of 78 aircraft were retrofitted, 36 at Embraer and 42 at PAMA-LS. On 20 March 1987 work on the first EMB-312 to be retrofitted at the Lagoa Santa facility, FAB 1370, was completed. The aircraft, together with other detail changes and modifications, had its operational cycle extended by 200 hours.

Also that month, the PAMA-LS workshops completed their first overhaul and maintenance on an EMB-312. This was FAB 1303, which was in fact the first airframe delivered by Embraer to the Air Force Academy back in 1983.

From the initial order of 118 EMB-312s placed by the FAB, the last aircraft was delivered on 14 August 1986 directly to PAMA-LS, sporting a Vietnam-style camouflage scheme and the FAB 1420 registration, before being delivered to its operational unit.

Finally, on 9 February 1990, the T-27 Tucano Hangar was formally inaugurated in a ceremony attended by Brigadier Octávio Júlio Moreira Lima, then Minister of Aeronautics.

Tools and ground support equipment were designed to provide the highest level of efficiency during EMB-312 fleet maintenance work. (PAMA-LS)

2ª ELO – The first tactical use of the Tucano in Brazil

From its conception, the Tucano was expected to be armed. The use of the four hardpoints for a tonne of conventional ordnance – bombs, unguided rockets and machine guns – would aid in the training of fighter and attack pilots, as well as provide air forces with a means of conducting attack, armed reconnaissance, escort and close air support missions. In some cases and situations, depending on the aircraft, the Tucano could also undertake interception and air defence missions.

After the Smoke Squadron and the Air Force Academy, the 2ª Esquadilha de Ligação e Observação (2ª ELO, 2nd Liaison and Observation Flight) was the next unit to operate the Tucano. The unit was the first to receive the aircraft in a Vietnam-style camouflage pattern (light grey on the lower surfaces, two shades of green and tan on the upper surfaces) and, in fact, was responsible for creating the entire attack doctrine for the Tucano.

The 2ª ELO, created in 1956, was responsible for supporting the Brazilian Navy, flying missions that incuded naval attack and ground attack and also assisting the ships of the fleet in the calibration of their radars and weapons systems. Due to this close connection with the Navy, the 2ª ELO was based at São Pedro da Aldeia Naval Air Base, in the north of the state of Rio de Janeiro, very close to the sea.

By the 1980s the FAB had entered a period in which attack aviation had been fully incorporated into fighter aviation. Attack missions were part of the operational profile of fighter squadrons and there was no single unit that specialised only in attack. Responsible for the coordination of these missions by operational fighter squadrons at that time was the Comando Aerotático (COMAT, Tactical Air Command).

Although exclusively supporting the Brazilian Navy, the 2ª ELO was the only COMAT unit to perform reconnaissance, escort, close air support, ground attack and low-altitude, level and dive bombing missions. However, the 2ª ELO did not have the status of an attack unit and was not recognised as such.

When it was confirmed that the 2ª ELO would receive the Tucano, on 29 September 1986, 11 pilots and 18 non-commissioned officers and sergeants took the three-week Tucano course with the Air Force Academy instructors. At the end of 1986, the first aircraft were delivered to São Pedro da Aldeia and later, in 1989, the Tucanos were assigned the AT (Attack/Training) designation given by the Flight's own personnel. The missions were also assigned the AT designation.

'In 1987 I had the privilege of introducing the armed Tucano with the FAB as soon as I returned to command that unique Air Force unit. Because I had previously served in the 2ª ELO and was teaching at the Air Force Academy, where the Tucano had recently started advanced training, I was assigned to command the Flight.

'In the first place we had to totally change the tactics of employment aiming at a more comprehensive operation, especially with regard to air-to-ground employment. Two officers from Operations were assigned to stay at Fortaleza with the 1°/4° GAV, where they collected information and adapted it to the reality of the Tucano, since that unit flew the AT-26 Xavante', recalled Colonel João José Pereira Pinto.

As it had at the Air Force Academy, when used in attack missions with weapons, the Tucano presented some problems. Some changes had to be made to fix the Tucano's sight as well as measurements that were necessary to match its use to the aircraft.

'We had another problem, more serious. The aircraft's weapon box was not compatible, meaning you could not use it. It did not ignite. Since we were in a hurry, I stopped at the 1º/10º GAV where the famous non-commissioned officer Arjona – an electronics specialist who was recognised for his ability to understand and solve problems – after a week of studying the circuit, turned the box into a 'super box'. Later the problem was corrected and the circuit was sent to Embraer so the next ones would not come with the same problem', commented Colonel Pereira Pinto.

In 1987 there was still no weapons employment doctrine for the Tucano. However, the arrival of motivated and experienced personnel made the deployment of the EMB-312 for attack missions quick, efficient and without incident.

One of these people was Lieutenant Wagner Farias da Rocha, who was assigned the position of Chief of the Ordnance Subsection. He had just completed a training course for Gunnery and Bombing Officers at CTA in 1988, winning first place.

'At that time [Wagner] *showed to Embraer's engineers that the T-27's TO-34 angle calculation tables were incorrect. TO-34 is the aircraft's manual for the weapons system. In this way, Embraer corrected the tables and, in possession of the new data, Wagner recalculated all the sight's angles for all air-to-ground employment modalities, which were all wrong. He also defined all parameters for employment of all modalities, except air-to-air, which was calculated by the 1° GDA in 1994. Then, the 2ª ELO went to Natal in order to carry out its first official campaign of air-to-ground employment at the Maxaranguape firing range. Doctrine development began there'*, explained Colonel Rinaldo Nery.

The weapons inventory assigned to the 2ª ELO comprised C-2 pods with one FN Herstal 7.62mm (0.3in) calibre machine gun with 250 rounds, plus LM 37/7 (seven

One of the weapons loads included C-2 pods with one FN Herstal MAG 58P 7.62mm (0.3in) machine gun and 250 rounds each and seven-tube Avibras 70/7M5A 70mm (2.75in) rocket pods. (Archive João Paulo Moralez)

37mm, 1.45in rockets), LM 70/4 (four 70mm, 2.75in rockets) and low-drag EQ-LMF 70/7 (seven 70mm rockets) rocket launchers. Although the LM 70/7 launchers (seven 70mm rockets) were available, due to their high aerodynamic drag these were not normally used by the 2ª ELO. Bombs used were the BAFG-120 low-drag and, for training, the 11kg (24lb) BEX-11. Old 250lb (117kg) AN-M57 bombs from World War II stocks were stored at the FAB War Material Park and were widely used by the 2ª ELO Tucanos.

'In 1988 we started to modify the Unit's Instruction and Operational Maintenance Programme, directing it to the training of attack pilots, wingmen and squadron leaders. We did nothing more than copy what fighter aviation did respecting the restrictions of the AT-27, because there was no point in reinventing the wheel. Unfortunately, fighter aviation did not look kindly on this 'intruder' in their business and many times they did not help. Flying reconnaissance and attack missions was easy due to the low speed 333km/h [180kts], *which allowed for easy visualisation of the targets on the ground, as well as the reduced turning radius that facilitated use. All this was allied with the good four-hour autonomy of a turboprop aircraft. The most effective configuration was two C-2 pods (for self-defence) and two LM 70/4 pods. We trained a lot of low-altitude navigation with simulated attacks'*, added Colonel Nery.

The doctrine began to be developed with tactical training including navigation at low level, penetration, tactical turns and 'scissors' manoeuvres. These consisted of two formations of two aircraft and the purpose was to change the direction of the flight without making any radio contact. The leader, who was in one of the pairs, started a turn to cross the other two aircraft, determining the new course to be followed.

Tactical navigation included flights at 60m (200ft) altitude and qualification for this type of manoeuvre included night flights. In this case the pilots would make a day flight to recognise the area and the missions were flown only with a full moon and while maintaining a distance of 91m (300ft) between each aircraft. The altitude chosen was always 304m (1,000ft) above the highest point of the terrain below.

At the end of 1988 the unit moved to the southern region of the country where it participated in Operation SULEX. Unleashed by COMAT, the 2ª ELO aircraft carried out armed reconnaissance missions, escort and use of AN-M57 bombs at the Saicã firing range. It was proof that the doctrine of attack had been developed and introduced to the Tucano in the correct way, since the exercise was a complete success.

'In the early 1990s the Air Force Commander of France came to Brazil to buy the Tucano. At the Santa Cruz Air Base some fictional targets were set up to be attacked by us, the Northrop F-5E and the Embraer AMX A-1. We were the first to attack because they thought that we were not going to hit anything, since the Tucano was relatively new in the squadron and they thought that we still had not mastered it. And the 2ª ELO destroyed everything. The pilots of the other squadrons got angry, but the Frenchman was delighted with what he saw. The attack was made in nosedive flight, something very difficult to do with the Tucano', explained Colonel Ricardo Magno Silva Rosa.

But even with all the attack doctrine under its belt, the main mission of the 2ª ELO continued to be support of the Brazilian Navy. The aircraft performed weekly missions for naval attack or alignment of the weapons systems of destroyers and frigates. The flights eventually included the participation of the aircraft carrier NAeL *Minas Gerais* (A11) and were even flown against ships of other friendly nations.

One of the main tasks of the 2ª ELO was to support the Brazilian Navy for calibration of radar and weapons systems. Here, two EMB-312s fly over the now retired aircraft carrier NAeL *Minas Gerais (A11)*. (Col João José Pereira Pinto)

'In this mission we were very strict because it was our benchmark. We knew every ship and its anti-aircraft defence system well. The simulated attacks on the ships of the fleet were 'Falklands style'. The ships were usually positioned in the 57km to 111km (40-60nm) stretch of coastline between Arraial do Cabo and Maricá. Some missions were performed, rarely, near Ilha Grande or Santos. In the area between Arraial and Maricá, the two aircraft took off from São Pedro da Aldeia and, after retracting gear and flaps, the wingman already assumed the cover position, while the leader manoeuvred to begin a low-level flight at the bow of the ship. As the ship approached, the wingman assumed the position of low-level penetration. Acquisition of the ship was visual, and as we approached, we were making the visual search on the horizon. The flight was very low, below the 91m (300ft) determined by the Air Command at that time. As we approached the ship, we were trying to attack from the bow or stern, trying to exploit the limitations of the ship's anti-aircraft system. The flight was low in order to escape the ship's radar. After the initial attack, radio contact with the ship was established, and flight profiles were met according to the orientation of the ship. Action time was between one and two hours, at most, and then we would return to São Pedro da Aldeia, in broken formation. On one occasion, I carried out an attack mission at 203km (110nm) of the coast where the targets were the Argentine frigates ARA Drummond *and ARA* Granville', said Colonel Nery.

Colonel Mario Sérgio Greskow Martinhão reported that for the radar calibration missions with the Navy the flight was made up of four aircraft up to 148km (80nm) or

185km (100nm) from the coast, close to the sea – three to four metres high – and with a very open formation with 300m (984ft) of separation between each other:

'We did not have radar and we followed the information that our intelligence had in hand or the radar data of the P-95 Bandeirulha marine patrol aircraft. The goal was to be detected as close as possible to the Task Force, though we realised that they detected us long before we made eye contact with them. Thus, when we started the manoeuvre to attack so they could work and prepare their defence, we were in practice already shot down because they already had us on the radars.

'So we started to fly very close to each other, close enough for the ship's weapon system to identify and lock on to destroy only one target – since the radar would only show one plot, one airplane. Then, when we got very close, we would break the formation and carry out the attack. The ship could hardly realign its system and reach more than one target. But as an aircraft was in sight, we knew that it would certainly be sacrificed in the attack. We were also attentive to other enemy aircraft.'

In a naval attack, the Tucano doctrine envisaged the use of 70mm (2.75in) SBAT rockets and penetration bombs. The 2ª ELO also supported the Corpo de Fuzileiros Navais (CFN, Marine Corps) in amphibious landing exercises and the beach-head conquest in Operation Dragon. The EMB-312 carried out the ground-attack and reconnaissance missions in a very complex military manoeuvre on the north coast of Santa Catarina. In the pre-planned attacks four aircraft were used, coming in from the sea to the mainland in a hi-lo-lo-hi profile and rocket attack.

On reconnaissance flights there would be two Tucanos moving at 304m (1,000ft) in a determined visual corridor, entering the operations area in a low-level penetration, coverage or front-line formation, depending on the topography and target to be hit. They used machine guns or rockets.

Weapons of the 2ª ELO included this low-drag, seven-tube EQ-LMF-70/7AP 70mm (2.75in) rocket pod. On the inner pylons are C-2 machine gun pods. (Luciano Porto, www.spotter.com.br)

Operation Anaconda

Around 12.00 on 26 February 1991, on the banks of the Traíra River in the Amazon and near Vila Bittencourt, on the Brazilian border with Colombia, 40 guerrillas of the Forças Armadas Revolucionárias Colombianas (FARC, Revolutionary Armed Forces of Colombia) crossed the water and attacked Army Detachment Traíra. They killed three soldiers and wounded nine others, from an effective force of 17 soldiers. All weapons, ammunition and communication radios were stolen by FARC elements, which in the same action also killed two illegal Colombian gold miners who had been detained by Brazilian soldiers.

As a consequence of this episode, the Brazilian Armed Forces, with the support of the Colombian military, unleashed Operation Traíra, whose objectives were to recover the stolen weapons and, above all, to show that the Brazilian military could deploy within a matter of hours in defence of national sovereignty. The mission was also a matter of retaliation and destruction of FARC drug dealers who exchanged cocaine for Brazilian gold.

The commander in charge of the Military Command of the Amazon was Army General Antenor de Santa Cruz Abreu, who headed the whole operation.

Within the FAB, the operation was codenamed Anaconda and saw the activation of three C-95 Bandeirantes, three C-115 Buffaloes, four UH-1Hs, one SC-95B Bandeirante (search and rescue), one RC-95 (reconnaissance) and four Tucanos of the 2ª ELO (registrations FAB 1391, 1398, 1400 and 1418). The Tucanos would perform armed reconnaissance, close air support and ground attack if necessary.

'*We went to a meeting at COMAT in Rio de Janeiro on 12 March 1991, and at 16.00 that day we received the Preparatory Order for the operation, which said that we had to prepare ourselves and wait for a possible start.*

'*The next day, from 18.30 to 20.00, we received training in the jungle and at 22.00 on 14 March the detachment was started*', Colonel Magno recalled.

A C-130 from the 1st Troop Transport Group took off from Afonsos Air Base to São Pedro da Aldeia to collect tools, weapon harmonisation systems, supplies, support personnel and mechanics. It took off at 11.15 towards Brasilia, from where it then departed for Manaus at 14.15.

At 10.00 on 15 March, four aircraft of the 2ª ELO took off from São Pedro da Aldeia to Alta Floresta and then to Manaus.

'*On arrival we had a meeting where they guided us and reiterated that one of the concerns was navigation in that region. The FAB squadrons used RADAM charts, a kind of aerial map obtained by the Federal Government in the 1970s to demarcate exactly the country's natural resources. It was something different, it was used for aeronautical navigation and was very difficult to interpret. The references were all visual. There was also a concern about in flight fuel consumption since the entire mission would be carried out almost at the limit of autonomy.*

'*It was decided that each aircraft would make two operational sorties per day. Maximum combat effort could reach four sorties per airplane daily. There was a warning on meteorology and orders to maintain radio silence during flights,*' said Colonel Magno.

Each mission would have between two and four aircraft with two pilots each, a requirement of the High Command so that navigation safety was maximised. Taking

A two-ship formation of EMB-312s prior to the transfer from Rio de Janeiro to Campo Grande.
(Carlos Filipe Operti)

off from São Gabriel da Cachoeira, it would involve a flight of almost two hours to the objective, 20 minutes over it and another two hours for the return. The mission was very long and on one of the flights the 2ª ELO entered Colombian airspace. The flight altitude was 1,524m (5,000ft), up to a distance of 180km (97nm) from the target, when the aircraft descended to near-treetop level to avoid being seen or detected by radar.

The strategy was to keep the region under constant patrol. To achieve this the aircraft flew to the objective in pairs and when one pair was leaving the area another was already arriving in the patrol region.

The aircraft weapons configuration consisted of two 7.62mm (0.3in) machine guns on the inner pylons for self-defence and two rocket launchers or bombs on the outer pylons for the attack.

'We had a serious problem with our health, since in that region it is common for people to suffer some kind of intestinal infection cause by water. On 22 March, four pilots were off the flight because of this and we came to the conclusion that it was better to only drink Coke and beer because it was industrialised and presented no risk of infection,' remembered Colonel Magno.

The operation's D-Day was 20 March 1991 The next day a reconnaissance mission was flown by four aircraft of the 2ª ELO.

'The weather was a problem for us because it rained all the time. The airplanes were in the open and we had problems with the sensors that informed us if the canopy was closed and locked. Other than that the Tucano did not present any problem. Upon arriving in that region we became aware of a phenomenon called aru, *a very dense fog that lies close to the ground, covering its entire surface. This happens*

as a result of high temperature and rains. With that we could not see the track and we could not, therefore, take off', said Colonel Magno.

The Intelligence service and the Army Special Operations Forces, in demarcating the targets in Colombia, found that FARC had placed women and children in their camps as human shields. For this reason, the 2ª ELO did not use any of the armaments, following a decision of the operation's high command.

In all, the 10 pilots and the four aircraft flew 155 hours (including the round trip to the Amazon) over six days. The maximum effort reached three operational sorties in one day.

The operation ended on 28 March 1991.

Change of doctrine and base

On 19 July 1995 the 2ª ELO left São Pedro da Aldeia and moved to Santa Cruz Air Base. At the new location, the unit ceased to support the Navy and gradually devoted itself to the development of a new operational profile including the training of its pilots for air-to-air combat and air combat patrol missions.

In the late 1990s the FAB was already planning to disband the attack units by turning them into fighter squadrons – a change that actually took place in September 2001, and which included the 2ª ELO. In the field of air-to-air combat, one of the main feats of the 2ª ELO was the development of a doctrine of dissimilar warfare against helicopters.

In the same decade, the air order of battle of the FAB tactical squadrons was directed towards the south of the country, mainly observing the border with Argentina. The doctrine and combat units emphasised possible combat or friction in that region involving the Argentinian government.

When envisaging the development of this dissimilar combat, the 2ª ELO sought support from the 5°/8° GAV 'Esquadrão Pantera', at the time equipped with the Bell UH-1H. However, that helicopter was already very old and would not have the desired performance to accomplish this mission. It was necessary to look for a model that was more manoeuvrable, smaller and agile.

'Our doctrine sector reviewed new techniques and combat missions and observed that helicopters had gained more prominence in the combat arena. We then identified that the 2°/8° GAV 'Poti Squadron', at that time based in Recife and equipped with the Helibras HB350 Esquilo armed with rockets and machine guns, was also willing to develop this type of doctrine. We created manuals, techniques and procedures to attack the helicopter without the possibility of a reaction. In the same way and in the opposite sense they sought to develop this doctrine. We campaigned to prove the development of these studies in practice.

'Thinking of an inverted cone, where the vortex is in the helicopter, our advantage was staying up and moving to the tip of this cone, so we had energy, speed and agility. The helicopter could not get up and if we went down – losing power – the helicopter could manoeuvre on its own axis and hit us. In simulated combat the helicopters took advantage several times', explained Colonel Greskow.

In January 2004 the 2ª ELO was transferred to Campo Grande Air Base and on the 30th of that month it was deactivated and its aircraft and military personnel absorbed by the 3°/3° GAV 'Esquadrão Flecha' ('Arrow Squadron').

1°/5° GAV: From the Academy to the formation of attack pilots

In the mid-1980s, while the AFA was responsible for training military pilots, the Centro de Aplicações Táticas e de Recompletamento de Equipagens (CATRE, Tactical Applications and Crew Readaptation Centre, 1 January 1974 until 1989), based in Natal, trained transport pilots for twin-engine aircraft, flying the C-95 Bandeirante in the 1°/5° GAV 'Esquadrão Rumba' ('Rumba Squadron'), and fighter and attack pilots in the 2°/5° GAV 'Esquadrão Joker' ('Joker Squadron') with the AT-26 Xavante.

The 5th Aviation Group was, in practice, the unit in which the newly formed military aviator from the Academy learned how to fly the aircraft tactically, exploiting all its capabilities to perform combat missions.

After the Tucano was introduced in the 2ª ELO – which in practice independently re-established FAB's attack aviation – the structure of the 5th Aviation Group began to change accordingly.

The FAB already planned to create new squadrons with Tucanos to counter illegal flights by small aircraft in the Amazon and border regions, reducing the drug and weapons traffic that had passed without major difficulties through a region that was previously poorly monitored.

Before long there would be a great demand for specialised pilots and with specific doctrine to fly the attack missions.

Thus, in May 1987, a T-27 Implementation Commission was created at CATRE and led by Major Zander Nogueira Martins. This provided for the formation of 10 initial flight instructors in 1988 and the beginning of the students' operational instruction in 1989.

This new task was absorbed by the 1°/5° GAV, which for some years would continue to provide instruction of future transport pilots using the C-95 Bandeirante and also administer the course for attack pilots in the AT-27 Tucano.

On 1 September 1987 the first 10 AT-27s were delivered, painted in a Vietnam-style tactical camouflage scheme. The aircraft arrived in formation with four C-95 Bandeirantes that performed several passes over the base before landing.

The 2ª ELO played a key role in this process of establishing the 1°/5° GAV. Through their pilots and weapons specialists, the experience, knowledge and doctrines devel-

With low-visibility squadron emblem, but high-visibility markings, this FAB 1392 from the 1°/5° GAV was seen in 1999. The heavy weathering was the result of frequent use for attack pilot training. (Col Aparecido Camazano Alamino)

oped by the 2ª ELO were passed on to the military of 'Esquadrão Rumba', which on 7 May 1988 bgean the Mutual Support Operation with the 2ª ELO. Here, four AT-27s from the 1°/5° GAV were used for the first time as weapon platforms.

Finally, on 25 April 1989, after having developed the entire course for attack pilots and successfully deployed the AT-27 at CATRE, the first class began its course. On December 5 of the following year, the Tactical Applications and Tactical Applications and Equipment Recomplement Centre became the Comando Aéreo de Treinamento (Air Training Command) and retained the same CATRE initials. The 1°/5° GAV, in turn, kept its structure divided into four flights – yellow, blue, white and green.

The first phase of operational training consisted of tactical training, where the students learned to fly farther from each other in offensive or defensive positions with two or four aircraft depending on the situation, such as tactical front line and low-level penetration. The next phase was ground attack with four aircraft at the Maxaranguape firing range, where aerial gunnery training, low-level bombing, level bombing and rocket launching were carried out.

For aerial gunnery the weapons employed were two 7.62mm (0.3in) machine gun pods and two BEX-11 exercise bombs, inert training stores mostly made of cast metal and teardrop-shaped to improve ballistics. They were usually painted blue, but some batches were painted in matt black. For low-level bombing missions two SBAT 37mm (1.45in) rocket pods were used. Finally, the next phase of aerial gunnery consisted of a T-27 towing a target, a 10m-long (33ft) by 2m-high (6ft) strip attached to a 100m-long (300ft) cable.

'One of the highlights of the course was the air combat phase. By now, we had more experience in the aircraft and we would learn to take it to the limit of its manoeuvrability. The mission consisted of two airplanes taking off, each with a student and instructor. After the initial exercises, where one student starts as a target and the other as an attacker and then alternating the positions, we went to the 'free manoeuvre', that is, we came face to face, always going to the right. When crossing with the other airplane, we began the manoeuvres in order to be able to position ourselves at '6 o'clock of the enemy' to make the fictitious shots and win the combat. Obviously no pilot wants to miss an air combat, so the airplane was pushed to the extreme and it responded accurately if you knew how to use it. It was at this stage that, once again, I was impressed by the Tucano. At that moment, with more flight experience and technical knowledge, I was able to really understand

The 1°/5° GAV 'Esquadrão Rumba' also flew aircraft in the air superiority grey finish. This is T-27 FAB 1356 with BEX-11 (11kg, 24lb) training bombs and seven-tube LM 37/7A SBAT 37mm (1.45in) rocket pods. (Col Aparecido Camazano Alamino)

what the capabilities of this airplane were. We could manoeuvre it to the limit of stall or G loads and it responded in a very smooth and controlled way. I learned how to manoeuvre the T-27 with the pre-stall alarm buzzing in my helmet's headset, since that was the only way to win a fight. The turn radius slowed to an extreme and the airplanes passed so close to each other that we could see the other pilot's clipboard. And you can be sure it's not aviator's tales. So much so that when he was already at the operational air unit, a commander who had just assumed and had come from the F-5E, almost arrested all the pilots of the squadron after making a combat flight in the Tucano and learning that we manoeuvreed so close and at the limit. Of course, after explaining to him that this was the correct way to fight in the T-27 and also by losing some air combats to the new lieutenants, this commander ended up accepting the situation, manoeuvring in the expected way and even winning some fights', said a FAB pilot who asked to remain unidentified.

The closing phase of the course in Natal was the CATREX exercise where live-fire attack missions were carried out. These consisted of a low-level navigation through the interior of Rio Grande do Norte carrying BINC-200 (209kg, 460lb) incendiary bombs and live AN-M57 general-purpose bombs. The objective was to reach the Maxaranguape firing range and attack fictional targets. Each student flew with an instructor since live weaponry was used.

Like all FAB units, the 1°/5° GAV missions were not restricted to aerial instruction. From 13 to 23 July 1993, 12 AT-27s and a C-95 of 'Esquadrão Rumba' were mobilised in Operation Canoas. The airfield used in the operation, located in the city of Oiapoque in the state of Amapá and near the border with French Guiana, was discreet and remote, had an asphalt runway and limited infrastructure. Just 2.7km (1.6 miles) separated it from the border.

The purpose of this operation was to establish the Quick Defence Force, under the command of Brigadier Éden de Oliveira Asvolinsque, then commander of CATRE.

The exercise was supported by the 3rd Brazilian Army Border Company, the 1°/1° GT (Transport Group) with the C-130, the 2°/10° GAV 'Esquadrão Pelicano' ('Pelican Squadron') for search and rescue missions, the Helibras CH-55 Esquilo helicopters of the 1°/8° GAV 'Esquadrão Falcão' ('Falcon Squadron'), the AS332M Super Pumas of

Very weathered T-27 FAB 1348 from the 1°/5° GAV with 'Vietnam' camouflage and two C-2 machine gun pods. (Col Aparecido Camazano Alamino)

the 3°/8° GAV 'Esquadrão Puma' ('Puma Squadron') and the 1st Communications and Control Group.

Later that year the C-95 Bandeirante was withdrawn from service with the unit, leaving only the AT-27 remaining in operation with the 'Esquadrão Rumba'.

The 1°/5° GAV was involved in the First FAB Attack Aviation Tournament, organised from 15 to 17 May 2000 in Natal. In addition to 'Esquadrão Rumba', the 1°/3° GAV 'Esquadrão Escorpião' ('Scorpion Squadron'), the 2°/3° 'Esquadrão Grifo' ('Griffin Squadron') and the 2ª ELO also participated. The air-to-air gunnery competition was conducted on the 16th with the 1°/3° GAV becoming the champion and the other units being tied in second place. From 17 to 18 August competitions were organised for low-level navigation, level bombing and air-to-ground gunnery. The winner of the tournament was the 2°/3° GAV.

By 2001 there was already a move within the FAB to disband attack aviation and once again incorporate its roster of missions into fighter aviation. Thus, there would no longer be a need to produce new attack pilots as of February 2002, when attack aviation was finally closed down.

The final AT-27 flight at CATRE took place on 17 December 2001. Days later the 24 AT-27s were distributed to the 1°/3° GAV, 2°/3° GAV, 2ª ELO, 1st Fighter Aviation Group, the 1st Air Defence Group and the 1°/14° GAV 'Esquadrão Pampa' ('Pampa Squadron').

In 15 years of operation, 442 attack pilots had been trained by the 1°/5° GAV. Also in December 2001, the 1°/5° GAV was transferred from Natal to Fortaleza where it returned to train transport, patrol and reconnaissance pilots, once again being equipped with the C-95 Bandeirante.

Tucano in FAB fighter squadrons

After the AT-27 – or A-27 – had been introduced into operational service using weapons – including bombs, rockets and machine guns – the Aeronautics Command began to examine the possibility of equipping front-line squadrons with the EMB-312.

In these units – two groups and one squadron – the Tucano would be responsible for a range of missions. Of these, the most important involved administrative duties: transporting officers to locations they were required at short notice; liaison and observation; navigation and instrument flight training; assisting in the formation of air defence controllers; surveillance missions and interception of small aircraft; and also to ensure that the flight hours of F-5E and Mirage IIIE pilots were maintained.

The FAB was in an era of cost cutting and the idea of using the Tucano to maintain a portion of the required flight hours or to complement the training of combat jet pilots would help in this new reality.

On 19 April 1990 the 1°/14° GAV 'Pampa Squadron', based at Canoas Air Base in the south of the country, received four T-27s with FAB registrations from 1439 to 1442. All of them were painted in the Vietnam-style camouflage scheme.

For the training of air defence controllers, the Tucano filled a fundamental gap and proved to be an ideal platform for this mission. Because it was slower and had a smaller turning radius, for example, it allowed students and future controllers to better understand and assimilate the interception and air defence stages compared to the F-5E and AT-26. In addition, these jets were more expensive in terms of flight time.

While it assisted in this training capacity, the Tucano also helped the 1°/14° GAV to more effectively control irregular flights and, in some cases, illegal flights in the border area in the south of the country.

These flights, made by propeller-driven and other light aircraft carrying smugglers, weapons and drugs, were difficult and sometimes impossible to intercept using the F-5E or the AT-26 squadrons based in that region.

In maintaining the flight proficiency of the fighter pilots, the Tucano was used for navigation, IFR flights and even for air-to-air gunnery. The operational cost was much lower, and it provided the unit with greater availability of aircraft while preserving the F-5 for live interception flights or more specific training.

Simultaneous with the arrival of the Tucano – armed with two C-2 machine gun pods with a 7.62mm FN MAG 58P under each wing – the interception mission began to occur very frequently. On some occasions the pilots of the 'Pampa Squadron' forced irregular aircraft to make landings for further investigation.

In 1997 the T-27s of the 1°/14° GAV also participated in a binational exercise with Argentina, in which they simulated aircraft entering the airspace of that country. The Argentinians had to identify the target, its location and send two IA-63 Pampa jet trainers to carry out required international procedures.

In 1999, at the meeting of 48 heads of state in Rio de Janeiro, where political and economic issues were discussed, the T-27s of the 1°/14° carried out surveillance and air policing in the no-fly zone. Until June 2009, when the Tucano stopped operating with the 1°/14° GAV, 13 aircraft passed through the squadron at different times, including FAB 1312, 1354, 1355, 1369, 1377, 1381, 1382, 1422, 1439, 1440, 1441 and 1442. The average availability in the inventory of the 'Pampa Squadron' was around three to four aircraft operating simultaneously.

For the 1st Air Defence Group, which at the time operated the Mirage IIIE/D, the first three T-27s arrived on 22 June 1990, in the same Vietnam-style camouflage. They were FAB 1437, 1438 and 1443 and performed missions similar to those described for the 1°/14° GAV.

Because the unit was far from any border, intercepting small aircraft was less frequent and the aircraft were more often used to help maintain the pilots' flight hours at a lower cost, as well as to provide training for navigation, instrument flying and air

Another EMB-312 used by the 1°/14° GAV 'Esquadrão Pampa' was FAB 1369, seen here in 2006 with the Centennial Flight of the 14-bis (1906-2006) logo on the tail. (Leandro Casella)

Left: T-27 FAB 1440 of the 1°/14° GAV 'Esquadrão Pampa' armed with a C-2 machine gun pod during a patrol over Rio de Janeiro.
(Col Sergio Barros de Oliveira)

Right: With front-line fighter squadrons the EMB-312 was used to intercept light aircraft on illegal flights. This is FAB 1440 from the 1°/14° GAV 'Esquadrão Pampa'.
(Col Sergio Barros de Oliveira)

defence controllers. Meanwhile, they were also involved in actual operations on the border strip, where they intercepted enemy aircraft.

Over the years, the unit's badges were applied to the T-27, including the Anápolis Air Base designation (AN) on the tailfin, the symbol of the 3rd Air Force Group (to which the 1° GDA is subordinate) and the jaguar (the unit's symbol) painted on the top of the vertical stabiliser.

In March 2009, by which time the group was already flying the Mirage 2000C/B, the Tucano stepped in and intercepted a single-engine EMB-712 Tupi. The aircraft, flying without a flight plan, was stolen from the Luziânia Flying Club, 100km (54nm) away from the group's headquarters and was manned by a man and his five-year-old daughter. The Tupi was approaching the limits of Brasília and would not respond to the calls of the controllers.

A Mirage 2000C was launched and headed to the area while maintaining air alert in flight over Brasilia, while the unarmed Tucano intercepted and escorted the aircraft.

The aircraft finally ran out of fuel and came down on the parking lot of a mall in Goiânia, killing the father and daughter.

Within the 1° GDA the Tucano was finally retired on 4 June 2009, the last two aircraft being FAB 1438 and FAB 1444.

In the 1st Fighter Group at Santa Cruz Air Base (Rio de Janeiro), several Tucanos passed through the unit for periods similar to those of the aforementioned 1° GDA and the 1°/14° GAV. Aircraft operated by the unit included serial numbers FAB 1319, 1376, 1377, 1380 and 1402.

Seen at Santa Cruz Air Base during Brazilian Fighter Day on 22 April 1995, FAB 1438 from the 1° GDA wears full squadron markings.
(Mário Roberto Vaz Carneiro)

7º Esquadrão de Emprego Tático e Adestramento (7 ETA, 7th Tactical Employment and Training Squadron)

For millions of Brazilians living in the south, southeast and even the northeast regions, it is extremely difficult to imagine the importance and wealth of the Legal Amazon – a region that encompasses nine of the 26 national states. This corresponds to more than 5.2 million km² (61.2 per cent of the national territory), in which little more than 24.4 million Brazilians live, just over 12.7 per cent of the Brazilian population. It is a region hostile to human presence, mostly covered by dense forests, with temperatures over 35°C (95°F) and humidity above 92 per cent.

On the other hand, much of its current importance stems from the abundant natural resources concentrated there. Its biodiversity is unique; in addition to mineral wealth and other resources available in huge quantities, it has the largest hydrographic basin in the world, with more than 12 per cent of the world's fresh water. This makes the forest eminently suitable for illicit actions such as smuggling of illegal weapons, animals and drugs, as well as biopiracy. Maintaining vigilance, especially in the border region, is a constant challenge, and difficult to execute effectively, even with the most advanced technological resources.

From French Guiana to southern Bolivia, there are more than 12,000km (6,500nm) of borders.

Since the 1950s, Brazil has sought to effectively populate, develop and integrate the Northern Region into the political-social context of the country. At the time, the Amazon was devoid of any highway linking its main cities to the greater urban centres of the midwest, such as Cuiabá and Brasília. Roads and access began to be opened in the Amazon, breaking the isolation of the region, giving more access to the residents and providing the armed forces with an opportunity to map and learn about the topography and geography of this inhospitable area.

From Boa Vista, one of the last five EMB-312s delivered by Embraer in 1993. This was part of the 15 aircraft with full provision for heavy armament, developed in the late 1980s. On the left side is the emblem of the 7th Tactical Employment and Training Squadron. (Col Aparecido Camazano Alamino)

Shortly after, in 1966, the Rondon Project was created, an initiative that began in the Brazilian Army and which led thousands of young college students to get to know places far from large urban centres, promoting social interaction with poor communities. In 1985, the federal government created the Calha Norte Project, whose intention was to militarily occupy the borders along French Guiana to Colombia, in a strip 160km (86nm) wide and 6,500km (3,500nm) long. In addition to the implementation of Border Battalions, creating a deterrent force and marking an effective military presence, the government also planned to populate this inhospitable region, creating ideal living conditions with infrastructure for schools, electricity, basic sanitation and other essential services to the population.

The FAB did its part by building air bases and hosting squadrons that could mark the presence of the force in the region and supporting army battalions established there as well. The 7° Esquadrão de Transporte Aéreo (7° ETA, 7th Air Transport Squadron) was created on 7 July 1983 in Manaus, equipped with C-95B Bandeirantes.

On 17 May 1984 the Porto Velho Air Base (BAPV) was activated, and this initially operated with a single Embraer EMB-810 Seneca. In mid-1984 the 7° ETA transferred its 2ª/7° ETA (2nd Flight), with C-95B aircraft.

Likewise, on 25 October 1984, the Boa Vista Air Base (BABV) was activated, consolidating its presence above the Equator Line and becoming the 1ª/7° ETA (1st Flight of the 7th ETA) also with the C-95B. In 1988, both flights began operating with the Cessna 208 Caravan, further enhancing FAB missions in the region.

Operation Anaconda in 1991 proved in practice the studies of the federal government and the FAB itself. The air arm intended to permanently place light attack aircraft in Porto Velho and Boa Vista, capable of rapidly intervening in any part of the north in a matter of a few hours, executing air patrol missions, attacking and intercepting small intruders and inhibiting illicit activities.

To consolidate these new policies, air transport squadrons were allowed to operate with attack aircraft such as the Tucano. This decision came about through Decree No. 768 of 10 March 1993, which also changed the 7° ETA's name to the Tactical Employment and Training Squadron – while retaining the ETA initials.

Soon, six AT-27s were sent to Boa Vista and Porto Velho, where they began to carry out their armed role.

'I was assigned to command the Infrastructure Squadron at Porto Velho Air Base, but I also flew in the unit as I was one of the four instructors and squadron leaders in the locality. The commander was a captain. The unit had only performed the air-to-air role under the supervision and instruction of officers from the 1st GDA. Its mission was airspace policing only', recalled Colonel Nery, who then went on to the 2°/3° GAV.

The 7° ETA was the only one of the seven tactical employment and training squadrons that actually received attack aircraft and, in this case, had the task of supporting the Sistema de Vigilância da Amazônia (SIVAM, Amazon Surveillance System).

However, that existence and the operation of armed aircraft was short-lived. In 1995, with the activation of the 3° GAV (3rd Aviation Group), the aircraft, pilots, mechanics and the entire structure of the 1st and 2nd Flights of the 7th ETA were transferred to this new FAB group. As a consequence, all the ETAs were once again employed exclusively for air transport.

Operations – the 3rd Aviation Group

The demise of the various flights of the 7° ETA (7th Tactical Employment and Training Squadron) marked, in fact, the consolidation of the presence and tactical role of the FAB in the strategic northern region of the country.

On 5 April 1995 the nucleus of the 1°/3° GAV 'Esquadrão Escorpião' and of the 2°/3° GAV 'Esquadrão Grifo' were created from the flights of the disbanded 7° ETA. At the time, it was decided that the new units would also fulfill ground-attack missions. Thus, officers from the 2ª ELO were responsible for all the instruction and implementation of this doctrine within the new squadron.

On 28 September 1995 the Aeronautics Command deactivated the nucleus and created the 1°/3° GAV and the 2°/3° GAV, at Boa Vista and Porto Velho Air Bases, respectively.

At the end of that year each flight doubled in size, and beagn to operate with six AT-27s per unit and an average availability of 70 per cent – the same level as found in the 2ª ELO. In 1999 there were eight aircraft in each unit and in 2002 the number rose to 12 Tucanos in each squadron. With the arrival of more aircraft, in 1996 the 1°/3° GAV started to operate with three flights – Scorpion Black, Scorpion Gray and Trump (dedicated to the commander and operations officer), but none had its exclusive aircraft. Within the 2°/3° GAV Squadron there were five flights names according to the suits of playing cards. These were Diamonds, Clubs, Hearts and Spades, plus the Trump, as also found in the 1°/3° GAV.

In those early days, the main challenges faced were the distance from the Lagoa Santa Aeronautical Material Park in the state of Minas Gerais, which provided support for the aircraft in terms of overhaul, maintenance, support and shipment of parts in case of incidents, accidents or breakdowns. The local facilities were incipient, lacking tooling and even qualified personnel.

'Gradually the mechanics experienced in the aircraft were passing on the knowledge necessary to the others. We were fortunate to receive excellent sergeants from the 2ª ELO and from the AFA, some of whom I had worked with. The high temperatures and humidity were not obstacles to the operation of the AT-27 in the Amazon', recalled Colonel Nery.

Surveillance operations

Until the implementation of the SIVAM project, which provided for the installation of a series of radars in the Amazon region to provide complete surveillance of Brazilian airspace, there were only radars in Manaus, Tabatinga and Boa Vista.ooling and even qualified personnel.

'To train interception, initially, we would go to Manaus. Afterwards, we started training in Tabatinga, supported by the 8° Batalhão de Infantaria de Selva [8° BIS, 8th Jungle Infantry Battalion]. *We carried out a joint operation, us and the 'Scorpions', from Cruzeiro do Sul in 1996, supported by the 62nd Jungle Infantry Battalion* [62° BIS]. *However, we did not intercept any aircraft. After this operation, we moved an aircraft to Cascavel, where we operated with other AT-27s from the first-line fighter units. It was the so-called Operation Sweep II; together with the*

Very weathered and with an uncommon 'fat' star, FAB 1393 from the 1°/3° GAV 'Esquadrão Escorpião' during an air patrol near the border. (1°/3° GAV)

Federal Police, Federal Revenue and Army we conducted air policing missions in order to restrain and intercept illegal traffic in the vicinity of the Brazil/Argentina/Paraguay/Uruguay borders, under coordination of the Brazilian Aerospace Defence Command (COMDABRA).

'At that time, I intercepted an Embraer EMB-820 Navajo, which had penetrated our airspace from Paraguay. The interception occurred at 1,981m (6,500ft), but as I appeared next to it for interrogation, it turned over me, making a 180° turn and started a descent. I returned to the position and carried out a warning shot, already descending. We reached the interior of Paraná at low altitude and I was already communicating with Air Defence through the link made by a P-95 Bandeirulha aircraft of the 2°/7° GAV 'Esquadrão Phoenix'. I carried out a new warning shot already in low altitude and I must have hit some cattle ahead of me. That year the destruction shot was not regulated yet, so I followed the Navajo until I reached the fuel minimum and returned to the retreat in Cascavel. The suspect aircraft returned to Paraguay. In this mission I used 60 rounds of 7.62mm. Other interceptions occurred with a better outcome, i.e. with the suspect aircraft landing for investigation', said Colonel Nery.

Also, from 3 to 15 November 1996, the 1°/3° GAV and the 2°/3° GAV, each with two aircraft and supported by another T-27 aircraft of the 1° GAVCA and 1° GDA, triggered Operation Tricolour I with the objective of countering illicit air traffic on the border

Left: BEX-11 (11kg, 24lb) exercise bombs with smoke cartridges inside to signal the impact on a training target. (1°/3° GAV)

Right: The C-2 pod with one FN Herstal MAG 58P 7.62mm (0.3in) machine gun with 250 rounds. Cocking was done outside the aircraft by ground crew. If there was an in-flight problem, the pilot had no system to re-cock from inside the cockpit. (1°/3° GAV)

with Peru and Colombia. The Federal Police, Federal Revenue and the 7th Jungle Infantry Battalion also participated in this operation.

With the Brazilian Navy, the 1°/3° GAV participated in Operation Ribeirex II in Corumbá, from 13 to 19 September 1997, flying missions including visual and armed reconnaissance of terrestrial and fluvial objectives.

The weapons employed by the 1°/3° GAV and 2°/3° GAV were the same as used by the 2ª ELO. The only difference was the introduction of the 200kg (440lb) BINC-200 incendiary bomb. Other stores included 7.62mm (0.3in) calibre MAG 58P machine guns with 250 cartridges each; or two 70/7 pods of 70mm (2.75in) rockets; or two Mk 81/BAFG-120 bombs of 119kg (262lb) each. Although they had four hard points (two under each wing), the Tucano rarely flew with mixed configurations, since this greatly increased fuel consumption. In some rapid-response operations one aircraft was armed with four BAFG-120s and the other with four rocket pods.

The SBAT 37mm (1.45in) rocket was rarely used due to lack of precision, and was employed mostly in training flights.

An important restriction was with regard to the dropping of bombs. These could not be launched sequentially, only simultaneously to prevent asymmetry that would cause aircraft instability. The Tucano can carry up to four napalm canisters provided they are equipped with stabilising fins. Without them, the bombs can tumble and collide with each other, hitting the aircraft. Without the fins, two napalm canisters can be dropped.

In terms of machine gun pods, the Tucano could technically carry up to four, however, due to wiring problems, the electrical system could not handle the firing of all at the same time and these stores were normally only used on the inner pylons. The problem was solved later for some foreign Tucano operators, who wanted to exploit the maximum capacity. The FAB also received a batch of 15 of these aircraft that were manufactured by replacing those that had been transferred urgently to Argentina. These 15 aircraft, being produced in 1990 (10 aircraft) and in 1993 (five), were delivered in the configuration permitting the use of more heavy weapons. All were delivered directly to the 3° GAV squadrons.

Another restriction on the 7.62mm C-2 pods meant the machine guns could not be cocked from inside the aircraft, only by the mechanic who pulled the charging handle before take-off. If the machine gun jammed in flight the pilot would lose that armament. For actual missions the rounds were mixed in a ratio of four normal rounds to one tracer. However, sighting the shots with this calibre was difficult during daytime. These two problems did not occur with the 12.7mm (0.5in) calibre pods.

Pilots from 1°/3° GAV 'Esquadrão Escorpião' prior to take-off from Estirão do Equador during an air defence alert deployment at the 4th Special Border Platoon from where they were headquartered. The aircraft had markings and registrations obscured. (Archive João Paulo Moralez)

In 1997 the two squadrons were the first units to hold air-to-ground training at the newly created Cachimbo firing range at Campo de Provas Brigadeiro Veloso (CPBV, Brigadier Veloso Proving Ground).

The squadrons also took part in the so-called Operation Highway, in which pilots had the chance to operate from federal highways. In such cases the unit as a whole could train its capacity for mobilisation. Infrastructure was minimal and the duration of the operation was a maximum of two days to avoid major inconveniences to local traffic.

'These operations are always preceded by coordination with agencies such as the Military Highway Police, the Federal Highway Police and also with support of ambulances and health personnel. When possible, an air traffic control structure is also set up. No type of armament is used, and the supplies are the minimum necessary to service', explained Colonel Nery.

In 1998, during Operation Mamoré, held in Guajará-Mirim and coordinated by the 17th Jungle Infantry Brigade, there was participation by the 4°, 6° and 54° BIS; also involved were the Special Forces Battalion and the 4th Brazilian Army Aviation Battalion (4º BAVEX).

This was the first army exercise employing the so-called Resistance Doctrine, with six 'Griffin' aircraft remaining in Guajará-Mirim to act as invaders. This doctrine aims to counter guerrilla actions, and develop strategies and tactics for irregular warfare including ambushes on jungle terrain.

On another occasion, four 'Griffin' officers were left in the jungle by a helicopter of the 4th BAVEX. The mission was to escape to Porto Velho while being pursued by a 54° Batalhão de Infanteria de Selva (BIS, Jungle Infantry Battalion) company.

'The great difference of the scenarios was the performance of the 2°/3° GAV and the 2ª ELO. In the Amazon, where things 'happen', the forces are more connected and interactive. Our connection with the army was very significant and close, since Porto Velho is the seat of the 17th Jungle Infantry Brigade. We also had very close contact with the 4th BAVEX based in Manaus. In 1996 all the pilots of the squadron carried out a jungle stage at the 54th Jungle Infantry Battalion in Humaitá. And we still had the air defence missions that the 2ª ELO did not perform at the time', concluded Colonel Nery.

Support for Special Border Platoons was provided in 1999, when air defence alerts were also conducted using an armed Tucano detached to border runways ready to intercept any illicit air traffic.

Operation Green Carpet – FARC in the sights of the FAB

'Upon being declared an Operational Attack Wingman I was assigned to serve on the 1°/3° GAV. Anxiety was very high for, despite being a small and distant city, serving in the Amazon generated a huge expectation since we were aware of the stories and missions that happened there,' recounted one pilot, an FAB officer who declined to identify himself.

'The first contact I had with the jungle was around May 1999, when the 1°/3° GAV and the 2°/3° GAV moved together to the Brigadier Velloso Proving Ground at the Cachimbo Ridge, in the south of Pará, to perform both air-to-ground and air-to-air

roles. The proving ground area is similar to Israel's and is heavily used for testing new armaments and the armed role of several squadrons.'

Over the course of 35 days, air-to-ground and air-to-air gunnery, low-level bombing, level bombing and rocket-launching missions were conducted.

The mission concluded with the use of real weapons and live ammunition. Again the aircraft would depart using low-altitude navigation to attack targets located in clearings in the middle of the jungle.

'We were solo on the aircraft and it was a matter of honour to find the objectives set for the mission. Without major problems we could locate the radar antennas and the fictional vehicles and throw our bombs on them. We made a pass on the target to see the damage and verified that accuracy was excellent – not much left for the next squadrons,' the FAB officer continued.

'In August I completed the last stage of the operational progression after making checks by instrument flight. The mission consisted of leaving Boa Vista, passing through Manaus, Cachimbo, Brasília, Belo Horizonte and São Paulo, carrying out instrument procedures in all these locations. The T-27 has great autonomy, but not such a good cruising speed. Another feature is that you need to do everything yourself, since it's an aircraft designed to fly solo most of the time.'

A few months later came the same pilot's baptism of fire, on the border with Colombia. When the mission occurred, only those who had the instrument rating could participate and this pilot was one of those sent into combat.

In the late 1990s, FARC was at its peak and dominated large areas of Colombian territory. In an effort to increase its domination, FARC succeeded in taking over the city of Mitu, in Colombia, capital of the Department of Vaupés and the main service and commerce centre of the province.

In this way, more than 1,300 rebels would soon control the entire region, killing 60 police officers and capturing some 70. Mitu is located in the far east of Colombia, just 40km (21.6nm) from the border with Brazil and the Querari Special Border Platoon (Pelotão Especial de Fronteira, PEF), in the state of Amazonas, part of the Brazilian Army.

Colombia could only rely on Apiay Air Base to launch an offensive that consisted of a force with two C-130s, two EMB-312s, three OV-10s, four UH-60 Black Hawks and three AC-47 gunships. Even then there would not be enough autonomy for the helicop-

A C-2 pod with one FN Herstal MAG 58P 7.62mm (0.3in) machine gun and 250 rounds was the standard weapon for air defence and interception of light aircraft on illegal or irregular flights. This is the Tucano flight line of the 1°/3° GAV 'Esquadrão Escorpião' in 2005.
(João Paulo Moralez)

ters. This was when Colombia discovered the existence of Brazilian Army runways near the border. One of them, Querari, was 40km (22nm) from Mitu and its use would be indispensable to the operation. While waiting for authorisation from the Brazilian government, an AC-47 was sent to discover the location of the runway.

The operation planned to retake Mitu with attacks coming from the west and east, in order to divide the rebel forces and facilitate entry into the locality. As FARC would never expect an attack from the Brazilian border, this was its least protected sector, and it was through it that Colombia's regular forces carried out incursions using armed helicopters and special forces, soon regaining control of the city.

The Brazilian government authorised the use of the Querari runway for 48 hours. Operation Angel Flight was launched on 1 November 1998. The deadline for the use of the airfield would expire in the morning of 3 November.

Flights by transport aircraft and helicopters began at night, infiltrating more than 240 special forcessoldiers who protected the runway and led the attack. With the use of flares, an AC-47 that was guiding the helicopter flight to Querari saw a column of guerrillas moving in the jungle, heading towards Mitu. The gunship was given permission to engage the guerrillas, resulting in the death of the vast majority.

'There is a video to which I had access, filmed by an army corporal, which shows the actions being carried out on the runway of Querari, where the soldiers appear disembarking after another raid – some of them were seriously wounded and others dead. In some scenes Black Hawks were targeted by large-calibre projectiles, but remained fully operational. It was also interesting to observe the logistics that they had set up on the runway to support the actions, with portable fuel tanks [7,570 litres, or 2,000 gal] *and MREs* [meal, ready to eat], *as well as weapons and ammunition. Subsequently, we learned that a Colombian Tucano aircraft armed with rockets and 12.7m* [0.5in] *machine guns also participated in the fight. Another remarkable scene is one that shows doctors and nurses of the Brazilian platoon attending the wounded, making sutures and field dressings so that the Colombians could return to combat.'*

Almost a year after this action, in which FARC had been defeated in Mitu and control returned to the regular government, army intelligence obtained information that the rebels would carry out another attack in Mitu. In order not to be taken by surprise again, they also planned to take the Border Platoon of Querari, since they believed

A Tucano from the 3°/3° GAV 'Esquadrão Flecha' during Operation Highway, when pilots and ground crew operated from federal highways to train their skills on short and narrow strips. (André Duailibi, www.spotter.com.br)

Displaying a very weathered air superiority camouflage, FAB 1392 from the 1°/3° GAV prepares to land at Boa Vista Air Base.
(1°/3° GAV)

that Brazil had supported the Colombian government in the past. In this way, the large-scale Operation Green Carpet was set in motion.

The Brazilian Air Force, Army and Navy all carried out specific missions showing force and presence in the region to demonstrate their intentions to the Colombian revolutionaries and disssuade them from attacking Brazilian troops.

'Thus, we were mobilised shortly before and on the scheduled day, 3 November, we took off with four Tucanos towards São Gabriel da Cachoeira, extreme northwestern Brazil. For the transfer the aircraft were without any tanks or external weaponry. A C-115 Buffalo gave us transport support, carrying our supplies, weapons, maintenance personnel and the rest of the pilots. Only when we arrived in São Gabriel did we have a notion of the operation's size. There was already the 2°/3° GAV, with four more T-27s, as well as some helicopter squadrons and several transport airplanes, many of them bringing live weaponry. AMX A-1 attack jets were also engaged in the mission, but remained in Manaus. We went to the initial briefing of the mission with the commander of the operation by the air force, a colonel of the General Command of Air Operations.'

There were more than 50 officers, including pilots, intelligence and maintenance officers, demonstrating the scale of the operation. The briefing presented the background history, the reasons for the operation and the planned actions during its duration. The T-27s left armed with bombs or rockets to attack targets very close to the border with Colombia, in the environs of the Querari Plateau.

'The stated objective of the attacks would be only a show of force to dissuade the FARC from their intentions, which, in fact, did not seem to be quite as expected, once

The head-on view from the rear seat reveals how easy it was for the pilot to see the student's actions or assist in an air-to-ground gunnery training mission.
(Archive João Paulo Moralez)

the actions unfolded. Since we had been instructed not to leave the premises or the airport, we used fictitious names and rank insignia glued to our overalls. We had to be very attentive during the flights, as we could be targeted by ground fire. Registrations were covered with black adhesive paper, as well as the base designations placed on the tailfin – BV or PV. The paint scheme was camouflage in two shades of grey and the registrations I flew in were FAB 1329, 1400 and 1446

'Following the briefing, which occurred earlier in the evening, the first attack orders were delivered the following morning, meaning the pilots scheduled for the first sortie spent the night planning the mission. Although I was only scheduled for the afternoon sortie, myself and all the other pilots helped to plan, primarily because we could not sleep early and, mainly, we wanted to be aware of every situation that was taking place and where the attacks would be, the line of action and the tactics that the most experienced pilots would use for the missions.'

Very early the next day, the pilots left their military shelters for the airport, which was about 12km (6.5nm) away. The pilots of the first sortie went directly to the squadron briefing while the aircraft were armed with bombs and rockets. Take-off was very early, an hour after sunrise, to avoid the *aru*, a white mist that covers the top of the trees; and soon after lunch, to avoid the characteristic late-afternoon showers.

'We did not have time to look at much because we soon received our attack orders for the afternoon and left immediately for planning. As we had seen the older pilots plan the first mission, we followed the same reasoning, but adapted it to the new targets that were determined. The flight leader distributed the functions for each pilot and we planned quickly, but checking all the times, approach and attack axes, drop

heights, bomb fragmentation envelopes and, mainly, the calculation of fuel, because we would have to take off and to land in São Gabriel da Cachoeira, for there were no alternatives on the route. I didn't even have lunch. I was so anxious, I ate whatever was around and waited for briefing, studying and memorising all the planning so as not to do anything stupid, since the whole thing was now very serious.

'We detailed each phase of the mission and, after clarifying all doubts, we set off for the airplanes. At that time anxiety subsides and the adrenaline is controlled because you only have time to think about each part of the mission and really focus on what you are doing so you do not make any mistakes. We started the engines and took off at the scheduled time. I was number 2 in the flight, armed with two bombs of the BAFG-120 type [120kg, 264lb, general-purpose bombs]. *Initial navigation was high, that is, we were high to save fuel, descending near the target area. Distance to the point of descent was approximately 330km* [180nm], *lasting one hour.'*

As doctrine stipulated, radio silence was total, with only the exchange of visual signals for checking aircraft conditions and fuel control.

The Tucanos of 'Scorpion' and 'Griffin' Squadrons were given GPS data to facilitate navigation, since there was little radio aid in that region, and a storm scope, which showed the position of the heaviest weather formations within a 35km (19nm) radius – something common in the Amazon.

'We reached the low-altitude flight approaching 90km [50nm] *from the target, proceeding to low-level penetration formation, and as we crossed the line of 'weapons ready' marked on the map, we switched the weapon panel on, preparing the stations. Dispersion of the aircraft occurred at 16km* [9nm] *from the target, each following its heading as planned, trying to remain visual with the aircraft in front of it. After three minutes of navigation, the leader broke radio silence and reported: 'Ace, PIB'. He was at the starting point of the climb. 'Ace in the dive,' he said. Then: 'Ace, bomb,' which meant the moment he had dropped the bombs. Thirty seconds later I reported my PIB and started the climb looking for the target area. Soon I visualised the wood and straw huts of a camp in the middle of the jungle, and the buildings on the right were my target. The ones on the left were already on fire because of the Ace bombs, from which a column of white smoke came out. Upon reaching the top of the ascent, I went into a nosedive, trying to keep the sight on the target, awaiting launch altitude. At the appropriate moment, I dropped the bombs, started the recovery, and spun the airplane to see the explosion. The bombs struck the front of two huts that exploded instantly, remaining on fire, while number 3 and number 4 also dropped their bombs.'*

After landing, the four pilots headed for the debriefing and filling in of the intelligence reports, where each one shared his impressions and what he had seen during the mission. This pilot flew two more missions in three days using basically the same flight profile as before. In one of the missions he launched a single ripple of 14 SBAT 70 rockets (70mm, 2.75in) at another camp on the bank of a river, well hidden among the trees. The machine guns were not used because they had only a penetrating effect and the objective was to cause the greatest possible damage. In some missions four bombs were dropped per aircraft, using all their weapons capacity. In other missions, 28 rockets were delivered per aircraft – seven rockets in each pod.

'We were not sure if the targets we attacked were camps or facilities already abandoned by the FARC or if they were still being used,' the pilot recalled.

The targets were over the border, with aircraft penetrating up to 3.7km (2nm) into Colombian airspace. Some pilots reported having observed certain movements but no Tucano was hit or attacked.

Army troops clashed with the Colombian guerrillas and after almost a week of positive results, demobilisation of the squadrons to their home bases started on 8 November. In the general debriefing all the results of the actions from the forces involved and objectives achieved were shown. There was no further evidence of any attempt to invade Brazilian territory and no further assault on the city of Mitu. The operation had been an absolute success and even months later there was no sign of a counter-offensive by FARC in that region.

Reports reflected the versatility, strength, precision and safety of the Tucano, the aircraft having suffered few breakdowns throughout the operation, and any problems were quickly remedied by the specialists.

'The T-27 was the main armed vector of an operation aimed at removing a real threat to our sovereignty and our people. The aircraft fulfilled its duty in an impeccable way, entering definitively into the history of Brazilian Air Force. I flew nine hours in five days, completing three missions of two hours each. The other flying hours were on the way to São Gabriel da Cachoeira.'

Other operational missions

The flexibility of the Tucano allowed flights to be made to the Special Border Platoons (PEF), units that at the time did not even have VHF radio. The Estirão do Equador airstrip on the border with Peru lies within forest and is 7km (3.77nm) away from the platoon's facilities. The pilots had to make flyovers warning of their arrival so that someone would pick them. A very low pass at reduced speed was made over the runway, looking for possible obstacles that could damage the aircraft or cause an accident.

After landing, the pilots were met by two soldiers armed with 7.62mm (0.3in) IMBEL FAL rifles, and equipped with a water jug and two hammocks. These soldiers, many of native origin, would be the sentries protecting the T-27 at night.

'The soldiers received us in the excellent standards of the Brazilian Army, with cries of 'SELVA' (jungle) *and full of excitement to see and understand the Tucano of the FAB. Entertained with the 9mm shots that we were firing from our pistols, which were obligatorily carried on this type of flight, soon we heard the noise of a tractor which finally appeared, pulling a bucket. In the engine compartment, placed there to keep warm, was a pressure cooker with food for the sentries that would guard of the airplanes. We started our 30-minute journey to the PEF, commanded by a 2nd Lieutenant of the Army.*

'The masonry and wood facilities were in good condition, with some water tanks covered with mosquito nets under the downspouts' outlets. The water on the Javari River was very muddy at that time of year; they needed to do that in order to collect clean water from the rain for consumption. The men of that PEF were rustic and professional, but humble even when facing the worst situations. They showed great respect and admiration for the FAB. In most of these platoons there is a phrase at the entrance that says 'From the first board to the last nail, this platoon was carried on the wings of the Brazilian Air Force'.'

These platoons also kept kerosene drums to supply the FAB's T-27s.

In the early 2000s, the 'Scorpion' Squadron was activated to reconnoitre for the 3rd PEF at Vila Bittencourt, near the border with Colombia, on the banks of the Caquetá River. The flight over the border focused on the point where the river leaves Brazilian territory and enters Colombia, the aircraft flying between 185km (100nm) to the south and 185km (100nm) to the north of the platoon. A few days before, there had been a confrontation between the Brazilian Army and a group of FARC guerrillas and reconnaissance was fundamental to check if there was any kind of movement in the region by the irregular Colombian forces.

Mission planned, take-off occurred early, with the T-27 armed with two 7.62mm MAGs under the wings, heading to São Gabriel da Cachoeira only for refueling. The flight plan was passed to the former Air Defence Cell of the Amazon and then the aircraft departed for the reconnaissance flight on the northern border of the Vila Bittencourt platoon. The target was reached after 40 minutes of flight and the search for signs of suspicious movements was initiated. After 30 minutes of patrolling, the aircraft landed at the 3rd PEF for lunch and supplies. The presence of a captain in the platoon led to suspicion, since the commander was usually a lieutenant.

'During the meal we started a conversation with the army officers about the mission and they started telling us about what happened. The river in question originates within Colombia, following an easterly course and, upon reaching the border with Brazil, makes a curve to the south, continues for approximately 35km [19nm] *over the border line and turns east again entering Brazilian territory. The 3rd PEF is very close to where the river curves into Brazil and serves as a checkpoint and surveillance for vessels that may try to enter our territory. On the Brazilian bank of the river, in the platoon area, there is a hut that serves as a control point of entry into Brazil, with military personnel and where the army remains on watch, along with a launch with an outboard motor and 0.5in machine gun mounted on its bow.'*

A 12.7mm (0.5in) armoured plate developed for the Tucano flown by the border squadrons at Porto Velho and Boa Vista. It was glued with double-sided tape and initially positioned further ahead and lower. Later it was repositioned as here. The system was not adopted and was only found on FAB 1396. (Mário Roberto Vaz Carneiro)

Two days earlier, at dusk, the Brazilian soldiers watched a boat slowly descending the river coming from Colombia, with the motor at low speed. The craft was very close to the margin facing the platoon, probably trying to go unnoticed. The vessel was sighted and received orders to land on the Brazilian bank to verify documentation and cargo.

There were five people on board, and when sighted, the boat accelerated. The army chased the suspect vessel with its boat, which ran aground. The Colombians, later identified as FARC members, opened fire but were shot and killed. No Brazilian soldier was injured.

After the episode, the Army Command was informed and positioned a considerable force in the region. The night after, a Brazilian military patrol spotted guerrillas in the middle of the jungle trying to get close to the platoon, but they were fired upon before reaching their target, fleeing into the woods. The FAB was alerted and then proceeded to send armed reconnaissance missions by the Tucano squadrons in the Amazon.

'After we heard this incredible story, the captain gave us a report that excited us because there was information that there were guerrilla encampments on the other side of the river, a few kilometres from the bank on the Colombian side. With that, he asked if it would be possible for us to enter Colombian airspace to try to verify such information.

'Obviously we had no authorisation for this, but after the story we had just heard, we did not think twice and said we would do the mission. We proceeded quickly to the airplane, checking the map of the region, verifying if the airstrip of Vila Bittencourt was headed to Colombia, which would facilitate our incursion at low altitude. We planned to ingress 10nm [18km] *into foreign territory, making a 180° turn and returning to Brazilian airspace in the shortest time possible, collecting whatever information was possible.*

'After the warm farewells, we set off and proceeded to the head of the river. We went through our planning and I, being in the front cockpit, started the take-off with the engine in full power. The Tucano quickly accelerated, leaving the ground well before the end of the track, but we remained as low as possible until we crossed the river. At that moment I climbed up to treetop level, which is not too low, since the jungle there easily reaches 50m [164ft] *in height.*

'Adrenaline and excitement were huge but we knew the risks we were taking, because if we were hit or something happened, we would not be supported by anything, possibly generating an international incident. Without a shadow of a doubt we were wrong, but we were new lieutenants with 'the knife in the mouth and blood in the eyes', as they say in the military, and that risk was what attracted us most.'

About 3.7km (2nm) after crossing the river the pilots saw something like a camp, with green tarps under the trees and a certain amount of of mess on the ground.

They continued straight ahead to avoid overrunning the camp and risking enemy fire. They continued until they reached the expected 18.5km (10nm), made a 180° turn and returned to the point where they saw the camp, marking their position with the GPS.

'We realised that there really was a camp there. Thank God, nothing abnormal happened and we came back safe and sound to Brazilian territory. We did a little more armed reconnaissance and, as the fuel was getting to the limit, we set the bow to São Gabriel da Cachoeira, going up to the most appropriate levels to

This six-ship formation demonstrates the variation of weathered colours and camouflage schemes among the 1°/3° GAV 'Esquadrão Escorpião'. (1°/3° GAV)

increase autonomy. When we arrived, we passed the data collected by HF radio from the ground station to the captain and the lieutenant of the 3rd PEF at Vila Bittencourt, who thanked us very much for our help and counted on us if there were any future engagements. Of course, these data were not included in our official report, because we had a full idea of what we had done and the consequences that would

A pilot stands alongside a seven-tube Avibras 70/7M5A 70mm (2.75in) rocket pod with anti-personnel warheads. Note the covered designation of the aircraft's air base on the tail. (Archive João Paulo Moralez)

have arisen. But more importantly, we had been once again involved in a real mission and were proud to be protecting our Amazon and our people.'

For the 2°/3° GAV, support flights to the PEF were common.

'Usually we received a Fragmentary Order [OFRAG] *that described the instructions for how this detachment would be, such as period, location, configuration and number of aircraft.*

'From there we prepared the mission. For the 2°/3° GAV aircraft, we usually went to Tabatinga, in the Amazon, where we were based. From there we made the trips to the platoons, having calculated beforehand the fuel necessary for the round trip, as well as the patrol time.

'There were two great advantages of operating the alert directly at the platoons. The first was the proximity of the routes used by illicit aircraft. These aircraft hardly ever entered Brazil near larger cities like Tabatinga and São Gabriel da Cachoeira. The second and biggest advantage was the secrecy regarding our take-offs, since the population of the platoons was much smaller and better controlled than in Tabatinga.

'Secrecy has always been a critical point to our operations. We had reports that our operations were monitored by drug dealers from Tabatinga and Letícia, a Colombian city bordering Tabatinga, which in practice makes up only one locality. Thus, we had the alert in Tabatinga and the control that was realised by Letícia, with people that we did not know. But even in Tabatinga there were 'lookouts' who knew what time our aircraft took off.'

Another point of discomfort for the pilots came at night, outside the alert period or when they were in the city of Tabatinga. Everyone had to be discreet because they could be identified and persecuted by drug dealers. The fear was heightened when Peruvian pilots who had a mission profile similar to the Brazilian's – including shooting down aircraft, something that was not undertaken by the FAB – were killed or threatened.

The pilots were called into action through two channels. When they were in the border platoons the satellite NERA telephone was used, since there was no rapid means of communication other than the HF radio. At Tabatinga the primary medium was the hot line, since the price of satellite connection at that time was high.

'After the call we would run to the aircraft in order to minimise reaction time and be as fast as possible in flight. There was a certain amount of pride on the part

Operation Cachimbo II at Brigadier Veloso Proving Ground in 1998. FAB 1339 from 2°/3° GAV 'Esquadrão Grifo' is configured with two BAFG-120 (Mk 81) bombs. (Col Rinaldo Nery)

of the pilots to be always faster, first to show the powers that were our competence and second to maximise the chances of interception.'

The 2°/3° GAV also supported the Brazilian Navy. The largest and most effective of these missions was Ribeirex, held in Corumbá, where two squadrons supported several units of the navy, including the Brazilian Marines. These were mostly reconnaissance flights without the use of cameras.

'This mission was really cool because we operated in a very real scenario, since the ships' role in relation to our aircraft was to kill us at all costs or to be sunk by us. After take-off, we assumed a slightly open formation (two aircraft) and descended, flying low over the Pantanal, literally at alligator's height, in order to make it harder for our aircraft to be seen by the enemy ships. When we found them, we prepared the attack, diving towards the target and simulating the firing of the weapon. At the same time the ship under attack attempted to kill us with simulations of Mistral infrared shoulder-launched missiles.

'In the major ships of each fleet we had a representative of our squadron, to judge the attacks and kills. These pilots were the first to go on a mission and stayed for up to a week aboard the ship without being able to go ashore. Meanwhile the components of the squadrons stayed in Corumbá, in an excellent hotel.'

On behalf of the Brazilian Army, the most common missions were those used to train forward air controllers, in which a soldier controlled the aircraft's strikes from the ground. Usually he was very close to the enemy troops being attacked, almost always in visual contact, which increased mission difficulty due to the risk of friendly fire. Instruction of these soldiers took place in the squadron itself, which provided the flights for the trainees' practice.

The 1°/3° GAV and 2°/3° GAV participated in the binational VENBRA II exercise held from 31 October to 6 November 2001, which involved the FAB and the Venezuelan Air Force and took place the city of Maracai (Venezuela). The exercise saw the participa-

A flight line with six armed Tucanos from the 3°/3° GAV 'Esquadrão Flecha' during Operation Highway. (André Duailibi, www.spotter.com.br)

tion of fighter, search and rescue and training units, and the missions included interception, attack and escort.

From 19 to 23 November 2001, Operation Combinex was carried out in several areas of the country, and the 1°/3° GAV was responsible for operating in São Gabriel da Cachoeira and Belém. The missions included ground attack, naval attack and interception.

From 2 to 7 December of the same year, Operation Presence in Fortaleza saw the accomplishment of attack missions in a strike package and joint employment with several FAB units, in the same format as a NATO exercise.

During Operation Japuru the FAB performed missions for territorial defence between 21 February and 4 March 2002 near São Gabriel da Cachoeira and Tabatinga, along the border with Colombia. Parachute troops were dropped along with supplies for the operation. The 1°/3° GAV carried out border patrol flights.

That same year other important operations took place, such as Tapuru, in conjunction with the navy and army from 9 to 16 June on the border with Colombia. The localities of São Gabriel da Cachoeira and Tabatinga served as headquarters for several FAB squadrons – including the 1°/3° GAV and 2°/3° GAV that performed visual and armed reconnaissance, escort and interception flights.

On 28 August, an Embraer E-99 airborne early warning and control (AEW&C) aircraft traveled to Boa Vista where it conducted a flight of 1h 10min during which it vectored a T-27 for interceptions at different altitudes, speeds and directions. The training was unprecedented within the FAB and was judged a success: the E-99 allowed the location of aircraft flying at low altitude close to vegetation and the forest – flight levels where detection by ground radars was impossible.

From 11 to 23 October, Operation Amazon I involved dissimilar air combat training (DACT) between the 1°/4° GAV 'Esquadrão Pacau' with the AT-26 and the 1°/3° GAV T-27, with 2-v-1 flights. Again, the E-99 participated in the control and in-flight coordination of these missions.

There were also joint exercises with the Brazilian Army Aviation. In 2002, in a show of force and presence at the border, the T-27s of the 1°/3° GAV and the 2°/3° GAV were stationed at São Gabriel da Cachoeira.

'One of these live actions consisted of a flight by two Tucanos – armed with two 70/7 rocket launchers each – to the 4th PEF Ipiranga on the border with Colombia for a mission with army helicopters. Once again, I was fortunate to be selected, along with three other pilots, to participate in this 'special operation'. We only knew the

On occasions, EMB-312s operated in conjunction with E-99 AEW&C aircraft. The Tucanos were vectored to search for illegal or irregular flights and force them to land. This is Tucano FAB 1387 of the 1°/3° GAV.
(1°/3° GAV)

The 3° GAV included the most active operational squadrons of the entire FAB. Operating close to the borders, Tucano squadrons like the 2°/3° GAV 'Esquadrão Grifo' were highly skilled and were involved in many operations. Seen in 1998, these two EMB-312s are armed with BINC-200 (200kg, 409lb) incendiary bombs. (Col Rinaldo Nery)

weapons configuration and that we would take off at dusk. The rest of the data would be relayed by the army when we arrived.'

While landing at the 4th PEF in Santa Rosa do Purus, the pilots saw two helicopters, a Black Hawk and a Eurocopter AS365K Panther from the 4th BAVEX parked in the platoon's courtyard. With them were 1st Special Forces Battalion soldiers who would participate in the operation to follow.

The mission consisted of taking over a clandestine gold prospection camp in the interior of the Amazon, within Brazilian territory, where miners were extracting high-value ores illegally, as well as destroying local fauna and flora. The camp was not far from the platoon, a 45min flight (at the speed of the helicopters that would transport the soldiers). The T-27 would provide escort with the possibility of becoming engaged if necessary and when requested by the troops on the ground.

Waking up early, the pilots got ready and went to breakfast, where they found all the army soldiers already equipped and ready for combat.

'At certain times their excitement was contagious. After the meal, we took all of our materials, reviewed all the details of the mission with the helicopter pilots and combat group commanders, and went to our briefing. The meeting did not take long, as everyone had participated in the planning and knew exactly what would be done. At the appointed time, we geared up the airplanes, prepared the weapons for the job and started the game.

'We took off and stayed vertical of the runway, around 304m [1,000ft] *high, waiting for the helicopters to take off. We saw special forces boarding the army aircraft, and then they took off, starting their navigation to the planned points. I was the wingman and we stood side by side behind the helicopters and higher than them. As*

we approached the helicopters, since we were faster, we made 'scissors' manoeuvres behind them, which consisted in the exchange of our aircraft's lateral position, placing a bow deflected 90° in relation to the axis of displacement, into the formation and then returning to the initial course after reaching the appropriate positions. The special forces' action happened very quickly and was a success.'

Another outstanding mission of the 2°/3° GAV was Operation Closed Gate, which consisted of armed reconnaissance flights within the limits of the Brigadier Veloso Proving Ground. The purpose was to prevent the routine invasions by loggers to steal timber from within the Proving Ground, surveillance being impossible to execute effectively only by land. The flight was made with one aircraft and two pilots, with the rear crewman carrying a camera to record the occurrences while a team followed by land or by river. During Operation Closed Gate from 14 to 28 November 2005, the 1°/3° GAV was able to carry out air defence missions in Belém with a single aircraft. The mission resulted in the seizure of 500kg (1,102lb) of cocaine, the arrest of a member of the local legislative body and more than 64 hours were flown.

The final squadron

The 3°/3° GAV 'Esquadrão Flecha' was the third and last flying unit to reinforce the FAB's presence in the Brazilian 'dry frontier'. The squadron inherited the pilots, mechanics and aircraft that had served the 2ª ELO, which in 2002 had been transformed into a fighter unit, divided into four flights: green, yellow, blue and silver. The latter was reserved for the commander and operations officer.

However, its headquarters remained far away from Santa Cruz Air Base, in Rio de Janeiro. It exchanged the temperate climate of Rio de Janeiro for the intense heat of Campo Grande, in the central region of the country. From there, the AT-27 was to intercept illegal or irregular flights coming mainly from the borders of Bolivia and Paraguay. This requirement received top priority from the Aeronautical High Command after the activation of new radars in the region identified a considerable volume of illegal air traffic. There was an urgent need to police the region completely.

The 'Arrow' Squadron began to take shape when it was activated in the form of a nucleus on 19 December 2003. From that moment on the aircraft and all the personnel of the 2ª ELO would be transferred to the new unit and gradually moved from Rio de Janeiro to Mato Grosso do Sul state.

On 11 February 2004 the nucleus was transformed into a squadron. In the same year, it was declared the most efficient unit of all the FAB fighter squadrons at the Fighter Aviation Tournament, a feat repeated in 2005. The unit, in fact, was constantly involved in deployments and training for attack, interception, escort, armed reconnaissance and combat patrol missions. Being a simple and easy-to-maintain aircraft, availability was always around 70 to 75 per cent.

As was the case in the 1°/3° GAV and in the 2°/3° GAV after they were transformed into fighter units in the early 2000s, an aircraft was always kept on alert, armed with two 7.62mm machine gun pods, fueled and ready for immediate activation. Attached to the inner side of the rear seat, on the canopy, a placard displayed to the aircraft intercepted the frequency 121.5MHz, as a way of establishing radio contact between the aircraft. The pilot was also armed with a 9mm pistol. The act of destroying a non-military

A three-ship formation over the city of Campo Grande by the 3°/3° GAV 'Esquadrão Flecha'. (André Duailibi, www.spotter.com.br)

aircraft was regulated in 2004, allowing air defence aircraft to shoot down aircraft that refused to comply with the recognised procedures for interception.

A highlight of the squadron was the possibility to participate with three aircraft in the multinational exercise CRUZEX in 2004. Here, pilots and mechanics came into contact with the doctrine and operational models employed in modern conflicts and theatres of operations in accordance with NATO standards.

On 12 May 2006 the first four A-29 Super Tucanos arrived with the squadron and the EMB-312 Tucano eventually ceased operating with the unit at the end of 2008.

The end of operational employment

Within the 1°/3° GAV and the 2°/3° GAV, as well as in the 3°/3° GAV, the EMB-312 Tucano ended operations in 2008, giving way to the A-29 Super Tucano.

Once it had been retired by these units and others that employed the T-27 in operational missions (such as the 1° GDA, 1°/14° GAV, 1° GAVCA), the remaining fleet of Tucanos was concentrated at the Lagoa Santa Aeronautical Material Park, in Minas Gerais, which until today is responsible for maintenance and major overhauls of the EMB-312.

With the end of the use of the Tucano with the DCTA in 2012 and its withdrawal by the Smoke Squadron in 2013, the AFA was left as the only FAB unit to fly the EMB-312, benefiting from the remaining aircraft of all these previous units. In some cases these aircraft did not immediately receive the standard white and orange colours characteristic of trainers, and could be seen on the AFA's flight line still sporting the operational aerial superiority camouflage in two shades of grey. Over time this camouflage gave way to the familiar training colours.

Today, no FAB EMB-312 Tucano retains any weapons capability. This occurred as the aircraft passed through heavy maintenance at the Lagoa Santa Aeronautical Material Park, which removed the sights, armament panels, cables and other components that allowed employment of weapons. These systems have been discarded and it is no longer feasible to adopt the FAB EMB-312 for combat missions.

It is worth mentioning that up to now, the Tucano has not undergone any modernisation with the FAB. The most critical systems are the landing gear. The FAB is studying the development of a new landing gear to replace the original and extend the service life of the aircraft. Another critical point are the engines, some of which need to be upgraded, while others need to be replaced entirely since they are now very outdated and worn out.

If nothing is done regarding the landing gear and engine, the EMB-312 fleet is likely to cease flying by around 2027.

But the FAB is already making efforts in this direction and also in replacing elements of the instrument panel and cockpit radios with newer, lighter models, which will be easier to maintain and operate. The cockpit is due to have one multifunction display in the centre of the panel with only two backup instruments displayed digitally. It is expected that the Tucano will remain in operation for another 20 years thanks to these modifications.

As of September 2017, no contract has been signed but the FAB was expected to receive its first 10 modernised airframes in 2018. With these upgrades, the fleet should stay in service until 2032.

Dusk at the Air Force Academy is not the end of the day for the EMB-312s, as the students embark on night flying training. (Lt Márcio Inforzatto)

SHORT BROTHERS S.312 – THE RAF'S TUCANO

Only two countries within the so-called First World operated the Tucano – France and the United Kingdom. In the event, France received modified aircraft with a longer operational lifecycle, French avionics and dive brakes. The Royal Air Force (RAF), meanwhile, employed an aircraft very different to that originally designed by Embraer.

In the early 1980s the RAF urgently sought a replacement for its BAC Jet Provost that was used for the basic training of its future military pilots. In addition to being obsolete for its role, the Jet Provost fleet was old, with low availability and very high operating costs – both in terms of spare parts and fuel.

The Tucano was already becoming well known in the international aeronautical world. It had taken part in the airshow at Le Bourget in 1981 and was at Farnborough the following year, where British military pilots had a chance to fly – and be impressed – by Tucano PP-ZDK, one of Embraer's prototypes and demonstrators.

And with the Tucano entering the international scene, the result would be a radical change in the RAF's doctrine of military training.

The Jet Provost was powered by a jet engine that consumed a lot of fuel and provided low efficiency in flight. In its favour, it was fast and manoeuvrable, with ejection seats and instructor and student sitting side by side.

The Tucano had a turboprop engine that consumed about one-third of the fuel of the Jet Provost's engine, steaed its crew in tandem, and although slower that the jet was much more manoeuvrable. In addition it was much more economical and easier to maintain than the Jet Provost.

In 1983 the RAF opened an international competition which called for the acquisition of up to 155 aircraft to replace the Jet Provost. The agreement would likely be valued at around GBP 150 million.

Air Staff Target 412 (AST.412), issued in August 1983, defined the requirements for the future RAF training aircraft. At maximum take-off weight, it was required to reach an altitude of 4,572m (15,000ft) in seven minutes from brake release on the runway. Operational ceiling was to be 7,620m (25,000ft), stall speed 111km/h (60kts), maximum speed 500km/h (270kts), and continuous speed at sea level 388km/h (210kts). At maximum weight it was required to use not more than 608m (2,000ft) of a paved runway to perform a take-off with a crosswind of 55km/h (30kts). The landing gear should be able to withstand an impact of 9.1m (13ft) per second during landing and the structure withstand G factor loads of +6/-3g. The seating configuration was to be tandem and the cockpit arrangement similar to that of the British Aerospace (BAe) Hawk advanced jet trainer.

As Embraer converted PP-ZTC to become a proof-of-concept for the UK Tucano programme, the airframe received a Garrett TPE-331-12B engine rated at 1,100hp and driving a four-blade propeller. It made its first flight on 14 March 1986. (Embraer)

The new trainer's missions were to include instrument flight training, day and night training and low-altitude navigation. There was no requirement for a specific type of powerplant. This, incidentally, was a key point as many in the RAF did not believe that the performance of a turboprop trainer could be comparable to that of a jet. The requirement meant that RAF technical staff could test and compare the qualities of different powerplants, ultimately forming a more concrete opinion and choosing the most desirable path to follow. The final decision would either maintain the traditional training model with the use of jet aircraft, or enter a new training model in line with global trends, especially in the less wealthy countries.

Of all the options that presented themselves – more than 20 at the time – only four were selected as finalists. Most manufacturers were rejected because they were not considered to offer a suitably high-quality product or they lacked the capacity to build an aircraft that would meet the demands of the project.

Of those chosen for the shortlist, all were turboprops. They comprised the Australian Aircraft Consortium (AAC) A.20 Wamira II; the ambitious NDN-1T Turbo-Firecracker project from the UK; the Pilatus PC-9, an improved variant of the PC-7; and the Embraer Tucano. None of the four options, however, was ready and none would meet the demanding RAF specifications without first undergoing modifications.

Winning the RAF competition would give the Embraer the chance to further project its brand identity and the Tucano itself, as the trainer would be operated by one of the world's leading air forces. Other nations would no doubt be attracted to the Tucano and might adopt it as their trainer.

As in every First World country that values its industry, each supplier was to partner with a national company to participate in the competition. The A.20 was the only entrant that was unable to find a domestic producer, while Westland committed to

The former PP-ZTC, here carrying the test registration G-14-007, applied as soon as the aircraft was ready for flight testing in Northern Ireland. (Archive Mário Vinagre)

assist in the production of the NDN-1T, Pilatus joined British Aerospace, and Embraer teamed with with Short Brothers.

There was, to some extent, scepticism within Embraer itself that it could compete with BAe in the United Kingdom. It was hard to believe that in the end the Tucano would have a real chance of competing and winning a contract of this magnitude. After all, Embraer had little more than 13 years of operational life at this time.

Confident or not, on 18 May 1984 a partnership between Short and Embraer was announced with the signing of a major trade agreement in London.

In the division of work between the companies, Embraer would be responsible for the necessary structural modifications of the design, tooling, and testing of these modifications. This package included development of an installation for the new engine (an item that was included later); modifications for the installation of new systems and internal changes in the cockpit to better resemble that of the Hawk; modifications to extend operational life; a ventral dive brake; structural reinforcement on the leading edges; necessary modifications to the main landing gear; in-flight engine and brake tests; testing of a new canopy; tooling development for the production of the new variant by Short; provision of complete structural kits (wing, fuselage and stabilisers) for assembly of the first 27 aircraft; main landing gears for all Short-built aircraft; and provision of the first prototype and 40 sets of structures and parts for canopies. In addition, the company was to provide training, technical assistance and documentation.

Short, in addition to integrating the British systems, would design the new canopy and produce the Tucano in Belfast, Northern Ireland, progressing from the assembly of ready-made kits shipped by Embraer to the production of several subassemblies and systems. The agreement also defined the markets in which Embraer and Short would act exclusively, and those that would not have exclusivity defined. In the latter, the responsible company would be selected once there was interest from a potential buyer.

At a meeting held on 24 April 1984, Embraer chose the airframe that would receive these modifications. It was PP-ZTC, which curiously – in a programme designed to meet a UK requirement – had the serial number 007 (312007), a happy coincidence with the legendary secret agent.

The company also defined at that meeting the changes that would be made to the airframe for the RAF programme. These included engine modifications that would allow the PT6A-25C/2 to deliver 850hp in exchange for a reduced lifespan; the dive brake; flap improvements to permit lower stall speed; a new bird-strike proof canopy and exhaust optimisation. Also included was a device on the wingtips to improve turning performance.

These changes would be paid for by Embraer. Work was to be completed on 27 July 1984 and was surrounded by the utmost secrecy. Everything would be done in a hangar with closed doors and there would be strict control and restricted access for the people who would work on the aircraft. It was forecast that 5,050 men/hours would be consumed for the modifications.

On 8 May 1984 the aircraft was taken to Hangar F-51 where it began to undergo changes associated with the power lever; the air conditioning system; the landing gear's normal control lever; improvement of the canopy seal; a new base for the control stick and the exhaust system.

On 31 May, the exhausts were cut back and reduced in order to further improve the Tucano's in-flight performance. However, it did not bring great aerodynamic gain, and ended up causing burns on the plating next to the engine that had to be replaced.

The engine arrived on 11 June 1984 and, in the next days, it was followed by its instruments, tools, mounts and other materials. On 20 June, the dive brake was installed and two days later the engine was started on the ground – separate to the aircraft – for the first time. On this occasion it did not reach the desired torque. On the 27th and 28th the engine was installed in the aircraft and it was first run on 2 July.

From 5 to 10 July, the aircraft was painted in a colour scheme similar to the other trainers used by the RAF – white and red. The next day the performance tests were completed. The wing and fuselage assembly was finished on 23 July.

Competition advances

The RAF eventually adapted the speed requirement to sea level, requiring that the maximum performance should be 444km/h (240kts) at full power and under the same conditions a minimum of 388km/h (210kts).

The first phase in the competition was an operational evaluation conducted using Embraer's PP-ZTT demonstrator from 13 to 24 August 1984 by members of the Air & Armament Experimental Establishment (A&AEE) at Boscombe Down. There was a total of nine flights comprising 20 hours. Ground trials were also conducted as well as a careful analysis of manuals and technical documentation to evaluate the aircraft's ability to comply with RAF requirements.

PP-ZTT made the Atlantic crossing together with the Smoke Squadron, which meanwhile toured Portugal and West Germany.

In parallel, Embraer sent PP-ZTC with the modifications to the UK. This aircraft would also be evaluated by A&AEE as the operational evaluation of the Tucano continued.

Before continuing with the assessments, the two aircraft were displayed on the ground and in flight during the 1984 Farnborough show. PP-ZTT, manned by Captain José Faria from the Smoke Squadron, and PP-ZTC, flown by Captain Celso Vilarinho, were responsible for representing the Tucano at this most traditional of British aeronautical events.

'Embraer wanted to sell the airplane to England and we went to the UK to demonstrate it. It was me and Vilarinho. And we kicked butt! In the time they gave us, which was very little, they offered an option. Three minutes for each airplane or eight minutes for the two together with an EMB-120 Brasilia, which would also participate in the show. We were captains and at the time we were very cocky. We accepted the proposal to do everything together,' recalled Colonel Faria.

'In the presentation that we trained for I would take off and, at 260km/h [140kts], *with the whole airplane shaking, I would pick up the landing gear and go inverted across Vilarinho. And from that began the demonstration with many close crossings, as we did in the Smoke Squadron.'*

The display was covered widely by the international media. It was even featured in the Brazilian national media, which traditionally neglected this type of event.

One of the witnesses to the presentation was the retired minister of the Superior Military Court, Dr Flavio Flores da Cunha Bierrenbach:

'The stupendous flight show of Major Vilarinho in 1984 with Embraer's T-27 Tucano in Farnborough – where, generally, everyone has seen everything and it is full of blasé people, who do not even turn their heads to look at a mere turboprop – left the audience enthralled and sealed the RAF's confidence in the Brazilian aircraft.'

After Farnborough, as planned, from 24 to 28 September, aircraft '007' flew to Boscombe Down to conclude a series of evaluations. Based on these results, the Tucano had the chance to meet the requirements if changes were made. After the show, PP-ZTT set off for a demonstration tour in the Middle East.

In the case of maximum speed at sea level, PP-ZTC reached 435km/h (235kts) instead of the 444km/h (240kts) required. It also failed to meet the requirement to reach 4,572m (15,000ft) in seven minutes from the start of the roll-out.

To attain the requested parameters, Embraer studied the adoption of a four-bladed propeller and the 950hp PT6A-62 engine, but this powerplant was already reserved for the exclusive use of Pilatus. Embraer opted to replace the engine with the Garrett TPE-331-12B developing 1,100hp. The Brazilian manufacturer offered the two engine options. If a speed of 444km/h (240kts) was mandatory, it would use the Garrett TPE-331-12B. Otherwise, the original configuration would be used, with the Pratt & Whitney modified to develop 850shp. Even if the Garrett was chosen, there would be technical support, maintenance and an operational guarantee with a pre-set annual cost value projecting an idea of the amount to be spent over several years. A partnership was also formed between Garrett and Rolls-Royce to manufacture and maintain engines in the UK.

Outside of the context of trials and evaluations, where the two competitors presented excellent results and came very close to the demands of the RAF, a political and commercial battle was initiated by the companies in order to seal the contract.

Short and Embraer evaluated the programme's costs at GBP 126 million and, on 21 March 1985, contrary to expectations, the Short/Embraer consortium was declared the winner. A total of 130 aircraft were initially ordered, with options for another 15. Short/Embraer's main advantage was in the total cost of the programme and it would also secure more jobs for the Northern Ireland economy that had been experiencing economic problems.

The contract was signed on 30 April 1985, and the aircraft was to attain a speed of 490km/h (265kts) at sea level at maximum continuous power. The use of the Garrett engine was mandatory in this case. The costs, project impact and programme risks with the new engine could not be detailed by Embraer and Short given the short time they had to assemble the final proposal for the RAF using the Garrett.

Development begins

Compared to the basic Tucano, the variant used by the RAF was to have a lifespan of 12,000 hours (against 6,000 of the Embraer Tucano); withstand G loads of +7/-3.6g; have a ventral aerodynamic brake; aerodynamic wingtips for improved turning performance; a new air-conditioning system; new, reinforced and easier to replace leading edges; a new canopy that was more resistant and with a completely redesigned layout; a new front fuselage section to house the new engine; a new cockpit layout; a new oxygen system, a flight data recorder and the use of avionics that were manufactured domestically. The basic Tucano's canopy resisted bird impacts only on take-off and approach to landing when speeds were lower. But the RAF wanted the canopy to withstand impacts during navigation sorties at low altitudes and at speeds in excess of 400km/h (215kts). For this, the canopy had to be divided and the front received the strongest reinforcement, as this would be the most likely to receive impacts. The ejection seat chosen was the Martin-Baker Mk 8LC.

There was a commonality of only 25 per cent of parts and components between Embraer's Tucano and the Shorts aircraft, and more than 900 modifications were necessary to meet the RAF requirements. The aircraft were, in practice, very different from each other.

The modifications on PP-ZTC began with the front section that was totally redesigned to accommodate the 1,100hp Garrett engine and the new Hartzell four-blade propeller.

Designated EMB-312G2 by Embraer, S.312 by Short and Tucano T.Mk 1 by the RAF, engine installation took place on 2 February 1986 and the maiden flight followed on 14 February under the command of Major Gilberto Pedrosa Schittini and engineer Luiz Fernando Oliveira. More than 14h 30min of flights were completed in Brazil until the prototype was dismantled and shipped to Belfast on 21 March 1986 aboard a Canadair CL-44.

After being reassembled and prepared for flight, the prototype with the test registration G-14-007 took off for the first time in the UK on 11 April under the command of Alan Deacon and Peter Rankin. Later the prototype received the British civil registration G-BTUC (British Tucano).

The Short Light Aircraft Company (Shorlac) was created as a subsidiary of Short Brothers to concentrate on the Tucano work. By the end of 1986 it was to assemble the

An air-to-air study of Kenya Air Force Short Tucano T.Mk 51 KAF 811.
(via Pit Weinert)

first two production airframes, carry out the testing campaigns and hand them over to the RAF to make its own operational assessment.

However, the first aircraft, registered ZF135, first flew on 30 December 1986 and was officially presented on 20 January 1987, at a ceremony held at the Shorlac facility in the presence of Ozílio Carlos da Silva, then president of Embraer; Sir Philip Foreman, president of Short; officers of the RAF and members of the British parliament.

The canopy was one of the reasons for the delays. The first canopy did not pass the test and the second example, with the transparent component further reinforced, did not guarantee correct and safe fragmentation so that the pilot could egress without problems in case of ejection. Therefore, a detonating cord was incorporated to aid in fragmentation as soon as the top of the ejection seat hit it.

In June 1987, ZF135 was delivered to A&AEE at Boscombe Down, followed by G-BTUC in October.

The results of Short's and the RAF's own trials saw some items undergo further modifications, which impacted in further delays. First deliveries were made to the operating units only in June 1988, including the Central Flying School at RAF Scampton.

In November 1989, Ferranti Computers Systems completed the construction of the first of five RAF Short Tucano full-motion simulators, comprising a front cabin and a system with three main projectors and a half-circle dome. Costing around GBP 1 million, the simulator was used in the initial phase before the cadet started flying the aircraft and allowed the instructor to make the first ground corrections of possible bad habits or mistakes by the students. In this way, some precious flight hours could be saved, reducing overall costs.

In January 1990 the UK Ministry of Defence requested improvements on the 50 airframes that had already been delivered, including revision of the communication

Wearing its desert camouflage, Kuwaiti Short Tucano T.Mk 52 KAF 106 is seen prior to delivery. (Archive João Paulo Moralez)

and navigation equipment, structural and flight commands. These changes would be incorporated into the aircraft that were still on the assembly lines.

The following month, Shorlac was the target of a terrorist bomb attack by the Irish Republican Army (IRA). Luckily an anonymous phone call allowed for the evacuation of workers, but one of the aircraft on the assembly line suffered considerable damage.

The final production airframe for the RAF was delivered on 25 January 1993.

Exports and final balance sheet

Embraer's association with Short Brothers proved advantageous for both companies. It was hoped that after the Tucano, other programmes would emerge in the field of commercial and regional aviation involving joint work between the companies. Clearly, it was hoped that the Tucano itself would bring many dividends. But it was not to be.

The way the contract was signed – at a fixed price, without passing extra development costs on to the British government – led to a drastic reduction in operational profits. And in fact, several modifications were requested beyond what had been predicted, all costs being absorbed by Short.

On the other hand, there was the expectation of leveraging profits with direct sales to several other countries. However, delays in the programme ended up favouring Pilatus with its PC-9, which went ahead by closing contracts with several countries. Production of the PC-9 and PC-9M exceeded 230 airframes, while the Short Tucano failed to exceed 160.

Embraer is estimated to have made about USD 35 million from the Short Tucano, but the main gain was as a consequence of having a military aircraft operated by one of the 10 largest and most important air forces in the world at the time.

Kenya was the first country to opt for the Short Tucano. However, the requirement was for an armed platform for weapons training flights and also to perform ground-attack missions. A total of 12 airframes were purchased in 1988, designated Tucano T.Mk 51. Short immediately began twork to install four hardpoints with a total capacity of 455kg (1,000lb), since the UK version did not have this provision.

On 11 October 1989 the first airframe for Kenya, provisionally registered as ZH203, made its inaugural flight. Short initially requested Embraer's support for the qualification to operate with external loads, including specific ground and in-flight tests. But Embraer responded that it would take a year to complete the entire programme, which led Short to opt for pursuing the work independently.

Tragically, during one of the high-speed vibration tests with external loads, ZH203 crashed into the sea near Rathlin Island, Northern Ireland. This was possibly the result of a structural failure due to extreme wing vibrations. Alan Deacon managed to eject, landing in the sea, but he died of hypothermia before he could be rescued.

Despite this setback, trials continued with live-fire weapons tests at FN Herstal in Belgium in April 1991, by which time deliveries to Kenya had started.

In February 1989, Kuwait approved the purchase of 16 aircraft, designated Tucano T.Mk 52. The inaugural flight of the first production airframe took place on 21 September 1990, the aircraft wearing the provisional registration ZH506. However, deliveries were only made in 1995 after the end of the 1991 Gulf War.

WORLDWIDE OPERATIONS

ANGOLA

Força Aérea Nacional Angolana (FANA)

Of all the countries that operated the EMB-312 Tucano, Angola was one of that most heavily users on ground-attack missions both in real operations and at war. In 1998 six aircraft were purchased and delivered to the country through a USD 15 million contract that included training and spare parts.

The aircraft, in fact, had been stocked at Embraer since 1996 and were part of a second order made by Paraguay covering six airframes. But the Latin-American country gave up buying and the planes were stored ever since.

Upon being acquired by Angola, the six EMB-312s were overhauled and registered from R-450 to R-455 and prepared for the three-step transfer flight, led by an EMB-120 Brasilia PT-ZBA. In the first one, which began on 26 November 1998, the planes left São José dos Campos, Embraer headquarters, to Fernando de Noronha, making a refueling scale in Recife. On the way out, however, one of the planes crashed into a fuel pump delaying take-off and forcing the crew to make an overnight stay in Recife before heading off to Fernando de Noronha. This first stage of the transfer was made with Embraer pilots in the front cockpit while the Angolans would remain in the rear position. From Fernando de Noronha the riders had another setback with a stop in the VOR system that had to be replaced. The planes took off but returned after identifying the problem.

EMB-312 R-451 taxis towards take-off for its delivery flight with two underwing 400-litre (105-gal) fuel tanks. (Mauro César Mezzacappa)

A new VOR was brought from São José dos Campos and only then the second stage was started on 28 November, continuing to Salt Island with the Embraer pilots still in the front cockpit. The third and last leg of the trip included the journey from Ilha do Sal (Salt Island) to Bamaco, Abdjam, São Tomé and then Luanda, in Angola. In this last section three of the planes had a mixed configuration of pilots (Brazilian and Angolan) and three with Angolan crew exclusively.

But on the take off from Salt Island caused another problem. One of the Angolan manned aircraft suffered a problem resulting in the forced landing of the R-451 with extensive damage in the aircraft, but without causing injuries to the pilots. It was then decided to keep Embraer pilots in charge of the rest of the five planes with Angolans accompanying in the rear seat.

The trip was completed on 3 December with the arrival of the five planes in Luanda. The EMB-120 PT-ZBA, which accompanied and served as support for the Tucano squadron, took to Salt Island on its return the Angolan aeronautical authorities that would make the investigations of the accident with the R-451. This aircraft was fully recovered at Embraer after having exchanged its left wing and other details on the fuselage. An Ilyushin Il-76 from the Angolan Air Force took the partially dismantled plane from São José dos Campos to Luanda and on arrival the Embraer teams reassembled it and returned it to flight condition.

The African nation was in fact the last customer to acquire Tucano. Initially its use would be focused primarily on advanced weapons training missions before the pilot moves on to operational units.

However, the reality of the civil war spread in the country did not allow the Tucano to be destined for instruction flights, but to concentrate the activities in the Esquadra de Reconhecimento e Ataque 401 (Reconnaissance and Attack Squadron 401) belonging to Luanda Air Base. Currently the 401st Squadron belongs to the Comando Operacional de Aviação de Transporte (COAvT, Operational Command of Transport Aviation) also at Luanda Air Base.

Civil War battles

In 1975, with Portugal having finally withdrawn its last military forces from Luanda and granting independence to that African country, Angola plunged into a long, bloody and destructive 27-year civil war with several stages over the course of nearly three decades.

In fact, just as in the Middle East, Africa presents a delicate balance of forces and a mosaic of events that, with the slightest outside interference, destabilize an entire nation in a short time. In 1992, countries such as the former Soviet Union, Cuba and South Africa, for example, had already deeply interfered in Angolan politics. In the country, the Movimento Popular de Libertação da Angola (MPLA, Popular Movement for the Liberation of Angola), which gained control of the country with help from Cuba and the former Soviet Union, in 1975; and the (UNITA, National Union for Total Independence of Angola), which initiated a movement of resistance and guerrilla war against the MPLA in this period.

But in the early 1990s the MPLA abandoned the ideology of Marxism-Leninism and a central party in the country and began to adopt a multi-party policy rather than a single one as previously ruled.

Thus, in 1992, elections that were intended to be free and democratic were organized in Angola with the participation of several parties. The MPLA won the presidential elections and UNITA went on to state openly that the election was neither fair nor free. UNITA had two-thirds of the territory of Angola under its command and thus incited a series of fighting in the country again generating a civil war.

In that atmosphere of confrontation, operational needs and demands for new military equipment and combat support, the EMB-312 fleet had already been widely used in counterinsurgency flights (COIN) attacking UNITA positions, reconnaissance and close air support.

From Luanda airplanes were deployed in detachments with four EMB-312s to the most varied air bases and distant aerodromes such as Saurimo Air Base, Catumbela Air Base, Lubango Air Base, Huambo Manoeuvre Aerodrome, Luena Aerodrome, Cabinda and Cuando Cubango Aerodrome.

Reports from authorities of the High Command of Angola highlighted the operation of Tucano as of great importance since the aircraft made possible and expanded the participation of aviation in the various missions to combat the internal enemy, having in some operations been the main weapon in the positive outcome that the fight had. Their participation was so positive that after those experiences in combat the Angolan state would choose to purchase the A-29 Super Tucano to continue this task.

The EMB-312 repeatedly faced enemy anti-aircraft artillery that sometimes hit the planes in flight, but none were effectively shot down.

Tucano's availability in the theaters of operations in Angola was high and the return of the airplanes to combat, whether for maintenance or damages suffered by the anti-aircraft artillery, was fast.

In those internal conflicts the country operated with 70mm (2.75in) SBAT 70 and Skyfire rocket launchers. General purpose bombs were also employed.

The first major operation with the EMB-312 was carried out in October 1999 in the well-known Battle of Bailundo with the airplanes flying from Catumbela in the Ben-

Left: Front cockpit from one of former Peruvian EMB-312s with gunsight. (Força Aérea Nacional de Angola)

Right: A very used and worn-out rear cockpit of a former Peruvian EMB-312. (Força Aérea Nacional de Angola)

Another training mission carried out over Angolan forest without any external stores. (Archive Pit Weinert)

guela province. The attacks were concentrated on UNITA's main facilities and headquarters causing a deep impact on the organization, impairing morale and the will to fight of its guerrillas.

In 1999, in the so-called Operation Zebra, the Tucanos participated in combat flights in the provinces of Cabinda, Zaire, Uige and Malange. The objectives of the mission were to free the city of Soyo and the military zones of Luzamba and Malange.

Four EMB-312 completed 25 reconnaissance flights, ground attack and attacks on both offensive and defensive forces deep into the Theater of Military Operations, always in line with the units involved.

The weaponry mainly used was the SBAT 70mm (2.75in) and Skyfire rocket launchers. In these operations, the aircraft were deployed at the Catumbela Air Base where the operational command post was established.

The aircraft departed from that aerodrome to the points previously indicated by the army troops as well as in the areas of aerial reconnaissance and attack to objectives on the ground. In general, the targets were groups of troops; fuel and ammunition depots; command posts; supply columns; communication centres and targets of opportunity.

The objectives of the operation were to re-establish the region's administration, to eliminate the instability foci as well as to open the communication and circulation routes.

In 2000, the EMB-312 participated in Operation Restoration operating provinces of Benguela and Huambo in order to liberate that area from the guerrillas. The armaments used, the targets attained, the objectives and the types of missions carried out were similar to those of Operation Zebra. However, this time, there were 35 flights in all, and the aircraft were posted at Saurimo Air Base (Lunda Sul province) and Luena Manoeuvre Aerodrome (Moxico province). The missions were carried out at the request of the units deployed at the military operations theater.

But the main and most important operation in which the EMB-312 participated involved the entire extension of the eastern military zone and the provinces of Moxico, Lunda Norte, and Lunda Sul. The objectives were to neutralize UNITA guerrillas.

Named Kissonde, the operation started at the end of 2001 and was extended until 2002 with four EMB-312s that made 30 flights on a secondment basis, covering the whole area of the operation.

In one of the missions, however, the airplanes located a column of UNITA troops and began guiding government troops to attack the guerrillas by maintaining the best possible attack position and gaining maximum efficiency. This particular flight culminated in the death of its leader, Dr. Jonas Savimbi. Thus, UNITA demobilized militarily, abandoned the armed struggle and concentrated its activities only on the political field. Once again the EMB-312 played a key role in ending the armed conflict.

More deliveries and other missions

The end of the armed conflict allowed Angola to widen the focus of the Tucano operation in the country.

Angola's plans for Embraer's advanced trainers included combating illegal mining in the northeast and centre of the territory where the diamond areas are concentrated; protect the eastern border that was the most violated; support the deployment of troops on specific missions and combat small armed groups.

In addition, the aircraft would serve as intermediate training for cadets. At the time, the training structure of the future Angolan military pilot consisted of the primary and basic stages in the Cessna 172K or Zlim Z-142 and only then they would move to EMB-312 with aerobatics flights, training, long distance navigation, visual and instrument night flights and use of rockets, bombs and machine guns. The airplanes were allocated to Escola Militar de Formação Aeronáutica de Lobito (Lobito Military School of Aeronautical Training) located in the province of Benguela after the civil war. The program included theoretical instruction and practice with approximately 100 flight hours.

The Tucano also served pilots, who already had a lot of experience flying aircraft such as the Pilatus PC-7 and PC-9, as an advanced training aircraft, for advanced weapons qualification and transition to high-performance jets such as the MiG-21, MiG-23, Su-17, Su-22 and Su-27.

Angola lost two EMB-312 during the operation, one of them being R 454. The accidents caused the loss of the airplanes' crew. It happened in the province of Cabinda due to technical failure and the second one was in Malanje, due to human error.

In 2002, in order to simultaneously recover losses, expand the fleet and the number of aircraft in operation, the country acquired six EMB-312 from Peru, stored at Las Palmas Air Base. The aircraft had the registraions FAP 425 (R-458), 427 (R-461), 436 (R-460), 437 (R-459), 444 (R-462) and 448 (R-463), respectively. All were out of operation, had been dropped from the inventory of Fuerza Aérea del Peru on 24 July 2001 and remained in storage since then.

Later that year, on 2 October, Peru's executive branch authorized the removal of the planes and allowed them to be offered for sale to countries that were interested. In 2002, Angola signed a contract valued at USD 4.8 million for aircraft that still had an average lifespan of 50 per cent to be used. FAP 448 was the newest airframe, delivered in 1992. The remainder was from initial batches delivered in 1987 without full capability, i.e. use of heavier armaments such as12.7mm (.50 in) FN Herstal HMP pods.

Although the negotiation was authorized, Peruvian justice embargoed the delivery of the aircraft to investigate possible irregularities in a sale of national patrimony. The investigations took two years to complete. Finally, the deliveries were made in 2004 and the aircraft entered operation in Angola only in 2005.

But in addition to Peru, in 2003, Angola purchased the two Embraer EMB-312 airframes that were used on demonstration flights at aerial shows, certification campaigns and aircraft presentation to potential customers, PP-ZTT and PT-ZTK. The aircraft underwent full overhaul, received the Vietnam-style camouflage already used by Angola and the registrations R-456 and R-457, respectively. The planes were delivered partially dismantled by chartered flight on a Brasmex Douglas DC-10.

The Tucano has already been withdrawn from aerial training in Lobito and is now concentrated in the 401 Squadron under Comando Operacional de Aviação de Transporte (Operational Command of Transport Aviation) in the capital Luanda.

Former Peruvian EMB-312 FAP 427, R-461 shows its weathered camouflage with shark mouth. Since the camouflage patterns do not match, the canopy was taken from another aircraft. (Lars Rohwer)

ARGENTINA

Fuerza Aérea Argentina (FAA)

In the early 1980s Fuerza Aérea Argentina (FAA) was beginning to suffer from the low availability of its advanced training airplanes Morane-Saulnier MS.760 Paris. The two Turboméca Marboré II engines of the Paris had high fuel consumption, were more polluting, emitted high noise levels and presented increasingly complex maintenance to be made. Getting spare parts was also getting harder. It did not have ejection seats and weapons capability was limited.

Argentina then began to seek options in the market to replace Paris in advanced training and the decision fell on the EMB-312 Tucano, after several studies and evaluations in flight made by Argentinian pilots.

The Argentinian Government bought 30 units that received the registrations E-101 to E-130. All were destined to Escuadrón II Tucano of the Grupo Aéreo Escuela (Aerial Group of School) of the Escuela de Aviación Militar (School of Military Aviation).

With the urgency of receiving the aircraft to place them in advanced instruction, Brazil ceded 15 aircraft that had already been formally delivered to FAB but had not yet been distributed to their operational units, in the case, the 2ª ELO. They all bore the Vietnam-style camouflage scheme. In this case, the only change made by the FAA was the application of orange paint on the wing tips and tailfin. Thus, Tucano was characterized as an aircraft used in instruction.

The aircraft were initially delivered in three batches of five aircraft. The first lot arrived in Argentina on 12 June 1987. The transfer was made with stops in Curitiba, Foz do Iguaçu, Cataratas and Reconquista. In August and September another two batches were delivered totaling 15 aircraft. The others were delivered in smaller batches as they left Embraer's production lines.

The traditional camouflage colours applied in Argentina for advanced training. (Santiago Rivas)

In Argentina the Tucano was inserted in the advanced instruction of the military cadet training cycle. Firstly, students flew the Beechcraft T-34B Mentor in primary and basic instruction for a total of 40 hours including the stages of pre-solo, training flight, aerobatics, daytime and visual navigation and a stage using a 7.62mm (.30in) caliber machine gun and practice bombs. Then they would go to the Tucano where they would fly another 80 hours in the stages of night flights, long distance navigation and using instruments.

Taking advantage of the ease of the airplane in aerobatic flights, Argentina decided to create an aerial demonstration team after attending an exhibition of FAB's Smoke Squadron. Thus, 20 pilots were selected and divided into the Yellow and Blue squadrons. Six aircraft were chosen and each squadron began training three times a week for three months.

In the first official presentation there was an accident at the moment when a Tucano touched the tailfin of another with its propeller in mid flight, pulling out its rudder. Both aircraft were able to land safely, the planes were repaired and the attempt to make an air demonstration team was immediately abandoned.

Although they received pylons to attach weapons, the FAA Tucanos never operated with any combat load. In fact, the country was the only one in Latin America that did not exploit these Tucano capabilities. In recent years the FAA studied the possibility of equipping airplanes with FN Herstal's HMP .50 FN machine-gun pods but the plan did not go forward.

In 30 years of operation, only four aircraft were lost in accidents, E-118 on 4 August 1988, E-127 on 20 December 1988, E-115 on 30 May 1990 and E-121 on 7 March 2006.

In 2009, Argentina suffered from the low availability of Embraer T-27 Tucano planes to continue advanced training of its cadets. Through a bilateral agreement, Brazil lent for a period of almost one year (31 July 2009 to 23 June 2010) five T-27 (FAB registrations 1319, 1344, 1369, 1380 and 1438) with a total of 1,500 hours available (close to 300 hours per plane). These aircraft received the Argentine registrations E-131, 132, 133, 134 and 135, respectively.

With the arrival of the T-6C+ Texan II, the Tucano will be used as a border control 'light fighter' to intercept illegal or irregular flights. It may also be used as a COIN aircraft. (Santiago Rivas)

EMB-312 E-110 shows it air superiority colours during a dive. (Andres Esteban Robles)

Preparations for transfer of the aircraft began on 26 July 2009 through the Plantax Mission established by FAX No. 36/SDTE/2674 and approved by Order No. R-998/GC4. The purpose of this mission was to provide technical support to the planes during the transfer and counted on a delegation of Brazilian mechanics who went to Córdoba, to the headquarters of the Escuela de Aviación Militar, on 26 July 2009. These mechanics had the task of passing to Argentinian technicians the differences between FAB and FAA planes and what should be the procedures adopted so that operation of the planes in that country was successful. Transfer of the aircraft itself began on 29 July by ten air force academy pilots (two pilots per plane). From Lagoa Santa, the planes stopped at Pirassununga and Santa Maria, then headed straight to Córdoba the next day, landing on Argentine soil at 17.50 local time.

At the end of 2014 the FAA chose to change the camouflage scheme of its Tucano fleet to the air superiority standard in two shades of grey. So far two aircraft have received the new paint scheme and the others will undergo this transformation after undergoing MRO – Maintenance, Repair and Operations. Currently around eight T-27s are flying in Argentina, however the FAA is in an effort to raise that number to 12 aircraft.

In August 2017, Argentina signed a contract to buy 12 Beechcraft AT-6 Texan II to substitute the Tucano at instruction at Córdoba. But current plans will move the EMB-312 fleet to another squadron and start intereception of light airplanes in illegal or irregular flights on borders. The fleet can also be used as COIN aircraft with FN Herstal HMP machine gun pods, seven-tube 70mm (2.75 inches) rocket pod, Mk 81 and Mk 82 bomb and BEX-11 practice bombs (11kg, 24lb) with smoke cartridges.

COLOMBIA

Fuerza Aérea Colombiana (FAC)

As the last Latin American country to purchase the EMB-312 Tucano, Fuerza Aérea Colombiana (FAC, Colombian Air Force) initially intended to use the airplanes in an advanced training program for its cadets replacing the already obsolete Lockheed T-33s with low operational readiness and no longer offering flight safety for that type of mission.

A contract signed in 1992 included 14 EMB-312, a flight simulator, spare parts and specific training for pilots and mechanics. Alongside the Peruvian airframes, Colombian aircraft were able to operate the full range of weapons including the .50 FN Herstal HMP machine guns pods.

At FAC the aircraft initially received a light grey colour pattern and high visibility insignia and badges, being registered from FAC 2250 to 2263. As early as December 1992 FAC received in two batches all 14 airframes ordered.

The first batch, initially with seven airplanes, left São José dos Campos on December 14 to complete the route to Apiay with stops in Brasilia, Alta Floresta, Manaus, Tefé and Tabatinga. But FAC 2251 had a problem in the navigation system and returned to São José dos Campos being delivered along with the second batch of aircraft that was then composed of eight airplanes and that left Brazil following the same route on 23 December.

The planes would be destined for the Escuela Militar de Aviación 'Marco Fidel Suarez' (Military School of Aviation 'Marco Fidel Suarez') to provide advanced train-

Colombia has not lost any of its EMB-312 fleet in accidents. Here, Tucanos FAC 2253 and 2257, from Escuadrón de Combate 212, are on the ramp. (Javier Franco Topper)

Within the FAC the EMB-312 played a key role in COIN missions against FARC guerrillas. This is FAC 2257 from Escuadrón de Combate 212 of Grupo de Combate 21. (Fuerza Aérea Colombiana)

ing with weapons. However, the war fought by the country against the FARC changed plans assigning them soon after to Escuadrón de Combate 212 (Fighter Squadron 212), of the Grupo de Combate 21 (Combat Group 21) subordinate to Comando Aéreo de Combate No. 2 (Combat Air Command No. 2) at Apiay Air Base. Also on site were the Rockwell OV-10 Bronco and FMA IA-58 Pucará.

The country soon developed an efficient and complete doctrine for Tucano operation, including use of heavy weaponry and maintaining an advanced readiness system that allowed its activation on a daily basis. Colombia, in fact, was the country that most used the Tucano in attack and CAS missions to its troops on the ground. In addition to these missions, the planes were involved in border patrol flights, intelligence gathering and interception of drug dealers.

From light grey and bearing markings in high visibility, the insignia changed to a low visibility paint scheme in two shades of darker grey. Around 70 per cent of the missions were carried out at night, a time at which the guerrillas were more vulnerable and also to disrupt the rest of the guerrillas. In 1997, the cockpits of all the aircraft were made compatible for operation with night vision goggles. This work had support from Peru, which already flew with NVG in its EMB-312 and thus provided technical support and instruction to Colombian pilots.

Although they were fighting guerrilla fighters almost daily and exposing themselves to the risk of ground fire, the airplanes did not receive any type of armour and no EMB-312 was lost in more than 95,000 hours flown since it entered active service in Colombia.

In almost all types of missions the T-27 were supported by the Cessna Citation SR-560 specially modified with sensors and radar able to search and locate targets in flight or on the ground. After location, information was relayed on so the Tucano could take action.

EMB-312 FAC 2257 armed with two Mk 81 bombs and two indiginous Indumil 56kg (125lb) conventional bombs. (Fuerza Aérea Colombiana)

At their headquarters in the Apiay Air Base aircraft remained fueled and armed with 15 minutes warning. Flights mostly involved two planes, each was equipped with a different type of weapon enabling a more effective attack on the guerrillas' positions on the ground.

Take-off was made using maximum engine power and the shortest length of runway possible to avoid enemy artillery fire and also lookouts that could provide early warning on departing aircraft.

Basically the Colombian EMB-312s flew on the attack missions with four launchers with seven 70mm (2.75 in) rockets each or two rocket launchers and two 12.7mm (.50 in) machine gun pods. The bombs used were always Mk 81 or Mk 82 in combination or carrying a single type on the aircraft's four hard points. On CAS missions, the aircraft carried two rocket launchers and two Mk 81 or Mk 82 bombs.

In missions triggered by emergency fire support against guerrilla positions, only two airplanes were usually deployed. As for pre-planned operations, attacks on targets already determined by the Colombian intelligence, it was common the involvement of four airplanes. However, in large operations the FAC came to use 10 EMB-312. In these cases, there was always a forward air controller positioned in a helicopter or AC-47 gunship that vectored the pilots for the attack. The controller could also be situated on the ground, being a national army officer, who also sent target indications for the attack.

Intelligence work was fundamental in all kinds of missions. Information including identification of the camp, its exact location, available weaponry, whether there was anti-aircraft defence, what was in the vicinity, escape routes and others. All data were passed on to the command that decided if those were in fact terrorists or not. If the information was validated the next step was the collateral damage analysis.

If the operation was approved, the responsible sector analyzed the number of aircraft to be used and the types of weapons ideal for that situation.

Once again the forward air controller participated recording the entire operation including the site of the attack and the air force's performance so that the collected data could be analyzed in the debriefing assessing the damage and if the established objectives had been achieved.

In 1995, the EMB-312 supported the 3ª Brigada del Ejército Nacional (3rd National Army Brigade) in attack and CAS missions. Comando Operativo No. 2 (currently Brigada 18 del Ejército Nacional - 18th National Army Brigade) also received an important fire support for the re-taking and control of the Departamento de Arauca (Arauca State) and preservation of the Caño Limón pipeline in Coveñas.

It was common for the unit to move airplanes in a secondment basis to other bases and airfields closer to the guerrilla and border region, as was the case of localities such as Tres Esquinas (Caquetá), Guabito (Cáli) and the Escuela de Aviación Militar (Military Aviation School) or Arauca and Palanquero Air Bases. This model of secondment allowed greater agility and speed in troop support since the FAC Command of Operations could activate the airplanes that were closer to the place to be attacked.

In this deployment system, using two airplanes, Comando Aéreo de Combate No. 3 located at Riohacha Air Base received in 1999 the aircraft that began to operate on combat flights. In that year an airplane at service of the drug dealers was shot down in Santa Martha. Several others were destroyed on the ground on clandestine runways. The following year, two other planes were deployed to Comando Aéreo de Combate No. 6, operating from Escuadrón de Combate 611 (611 Fighter Squadron) in Tres Esquinas to support a region critical to combat the FARC at that time.

On 20 February 2002, the Colombian EMB-312s actively participated in Operation Thanatos to retake a demilitarized area of more than 42,000km^2 (16,216 sq miles) in the south of the country that had been seized by FARC insurgents. Nearly all of FAC's combat aviation, including Mirage 5COAM, IAI Kfir C7 and Cessna A-37 Dragonfly jets, plus the EMB-312, OV-10 Bronco, Basler BT-67 and attack helicopters poured bombs and rockets on the region, mainly in San Vicente del Caguán. The whole operation was a success.

But in addition to combat missions the EMB-312 also perform the task of operational conversion in advanced training to train FAC combat pilots. The training includes use of air-to-air and air-to-ground weapons, tactical navigation, interception modes and attack profiles. More than 400 pilots have been trained to this day.

With the arrival of the first EMB-314 Super Tucano in 2006, out of a total of 25 acquired by the country, the EMB-312 Tucano were gradually withdrawn from the attack front line units and concentrated on the operational conversion missions of pilots entering combat aviation in the country. Currently the fleet is concentrated in two units, Escuadrón de Combate 212 and Escuadrón de Combate 611.

Second life

The importance of Tucano to FAC was of such magnitude that the country decided to invest in a process of modernization and revitalization of the 14 airframe acquired in 1992. The fleet reached 50,000 flight hours and some of the airframes had more than 10,000 landings. Many of the aircraft had already reached or were close to reaching 6,000 flight hours, the limit of their structural life.

Thus, a study was initiated and it was concluded that the modernization would be done with the replacement of communication and navigation systems; digitalization of the cockpit while maintaining the ability to operate with NVGs; replacement of around 25 analogue instruments with the installation of two, 9-in coloured multifunction liquid

Left: A new digital cockpit and head-up display will help to convert pilots to the Super Tucano and other modern aircraft.
(Javier Franco Topper)

Right: Front and rear cockpit with new panels.
(Javier Franco Topper)

crystal displays in each cockpit; installation of head-up displays; automatic pilot and complete replacement of wings and landing gear and some structural reinforcements in other parts of the fuselage. Although there were no new weapons systems compatible, the aircraft maintained the capacity to use the weapons already used by the EMB-312 in Colombia until then.

The process was led by Corporación de la Industria Aeronáutica Colombiana SA (CIAC, Society of Aeronautical Industry in Colombia) which has its headquarters in Bogota. As part of the offset agreement for the purchase of 25 EMB-314 Super Tucano, Embraer technically supported the Colombian program also trained 20 engineers and technical groups. The country has also become the only one homologated and certified to carry out the modernization of all EMB-312s in operation around the world with the exception of the airframes operated by Brazil.

Work was done at the headquarters of FAC's Comando Aéreo de Mantenimiento, at the Justino Mariño Cuesto Air Base in Madrid Cundinamarca. Modernization and revitalization of each aircraft took about four months and allowed the units to gain an additional 20 years of operational life or 6,000 flight hours. FAC will have a better performing, more reliable platform with lower operating costs, having replaced all systems that were in an advanced process of obsolescence.

The modernization was completed in 2016 with Tucano acting on advanced training missions to form future FAC combat pilots with air-to-air, air-to-ground, interception, attack and simulation of on-board radar operation doctrines, among others.

EGYPT

Al Quwwat al Jawwiya al Misriya (Egyptian Air Force – EAF)

Egypt was one of Embraer's largest and most important partners in the EMB-312 program, alongside the UK. After all, it was with the Tucano in Egypt that Embraer established its first assembly and manufacturing line under licence outside Brazil. The agreement secured two more military operators for its turboprop and 134 units produced. It was certainly a big deal for Brazil and to Egypt, as well.

In 1975, alongside Qatar, Saudi Arabia and the United Arab Emirates, Egypt led the foundation of the Arab Organization for Industrialization (AOI), which aimed to create and strengthen a high-tech military industrial base seeking the independence of these Arab countries from suppliers like France, UK, USA and the former Soviet Union.

AOI would be the catalyst for high-tech knowledge and would foster the development of the local economy of these countries by generating jobs and skilled labour by founding a set of defence industries. These industries, in turn, would develop technologically and aggregate expertise through agreements with foreign manufacturers when Egypt or AOI member countries purchased their products. Thus, these foreign manufacturers would commit to transfer technologies and to train their personnel, including licence production in Egypt at AOI headquarters.

Over the years these industries would become more empowered and could in the future begin the complete development of their own products in a completely independent manner. That was the idea until then.

And so it was done with Embraer.

Egypt was interested in the Tucano when it was still in its development phase in 1981. The Egyptian government called the Brazilian manufacturer to participate in a competition that could result in the acquisition of training aircraft by Egypt and possibly other Arab nations. The Pilatus PC-7, being in operation and already being a proven

An Egyptian Air Force EMB-312 painted and undergoing final assembly at Helwan before starting its flight tests. (Archive João Paulo Moralez)

product was indicated as the most viable option, while the Tucano was still only a prototype.

However, negotiations for the PC-7 acquisition based on licence production took a long time giving Tucano room to go into service and showcase its qualities and flight characteristics.

Thus, the contract with Embraer, valued at USD 181 million was signed in December 1983. It was the largest agreement that Embraer had signed until then. This included 40 EMB-312 for Egypt and 80 for Iraq, in a cooperation program divided into three stages.

The first included the training of the Egyptian workforce in Brazil and the production of 10 EMB-312 by a mixed team of Brazilians and Egyptians. The completion of these works would certify the qualification of the Egyptian labour force for continuation of the assembly of the remaining 110 aircraft in Egypt.

The second stage included the supply by Embraer of 42 kits of the great subassemblies and parts of the Tucano as sections of the fuselage, wings, tailfin, landing gear and others. These kits were loaded into aircraft and taken to Egypt where they were assembled by Arab engineers and technicians with Brazilian assistance. Finally, the remaining 68 aircraft, comprising the third and final stage, would be assembled in Egypt from parts supplied by Embraer.

Production begins

In Egypt, AOI had selected three industries to assemble the Embraer aircraft. The first, the Kader Factory, located in the suburbs of Heliopolis, had an assembly rate of two EMB-312s per month through the subassemblies, then the parts that would be received from Embraer.

In addition, Engine Factory, located in Helwan, licensed the Tucano's Pratt & Whitney engines, which received the designation PT6A-25E. The Eletronic Factory, also of Heliopolis, assembled some of the Tucano's electronic components.

It is noteworthy that the Kader Factory, for example, was very rudimentary, had a reduced staff, few tools and dirt floor. There was no more developed section within that industrial unit. Kader Factory professionals also did not have a high degree of specific knowledge in aviation.

In relation to the configuration of the aircraft, they were very similar to the Brazilian ones. There was only difference in the avionics that were simpler and that were replaced by French suppliers, as was the case of V/UHF radios. Embraer then designed all the necessary changes to install the production line and built the Tucano tooling for the Kader Factory. Work began in mid-1984 and took six months for all production areas to be structured including warehouses and supplies. There were 12 assembly stations until the aircraft was ready for test flights.

Already in April 1984, the first Arab technicians and engineers arrived at Embraer and underwent a theoretical course with accompaniment and work on the 10 airframes that were produced and assembled in São José dos Campos. The first production airframe came out of the assembly template on 27 June, and the final construction of the first EMB-312 of that contract, c/n 312085, began in July and only left the assembly line on 10 September. All 10 planes made in Sao José dos Campos were destined to Iraq. In December, the first four EMB-312s took off from Embraer for their delivery flight to Egypt.

While the work was coordinated in Brazil, in January 1985 two transport planes brought the Tucano assembly kits to Egypt. With the Egyptian technicians trained in Brazil began the production of the aircraft in Kader Factory. As they passed the 12 workstations and the Tucano was taking shape, the Arabs were learning more with the Brazilian airplane and the Embraer personnel absorbed the Arab and Egyptian culture. It was a job that worked very well.

But there was a peculiar situation with the Kader Factory. Years before, the factory lost its runway that was transformed into a avenue baptized as Tayaram (airport). Thus, when it was ready, the airplane could not pass through flight trials. It was an unusual situation, because all aircraft factories had their own runways.

It was necessary to transfer the Tucano from Kader Factory to Helwan, who had an airstrip. But it was 40km (21nm) separating the two facilities through the busy streets and avenues of Cairo and then following a path parallel to the Nile river. Once again the Embraer technicians took action and created two supports, one for the wings and one for the fuselage, using the suspension of an old Ford Galaxie. The plane was partially dismantled, placed in the wagons, transported to Helwan. Without yet having any paint, the wings and fuselage were separately placed in a small painting booth of Arab-British Helicopter where they were painted. Then the plane was assembled, tested on ground and then in flight.

In the middle of the Egyptian summer the Tucano with serial number 312105 made a flight of 40 minutes under the command of General of the Reserve of the Egyptian Air Force Gamal Farghaly. It was the first Tucano ever assembled in Egypt and to be destinated to Iraq.

Egypt also acquired two flight simulators and a few years later began to experience problems in continuing to pay the contract to Embraer. This situation was solved later. Together with the Brazilian government, the installation of a Tucano permanent assembly line in Egypt was scheduled to be installed at Helwan's Aircraft Factory. The project did not go forward, but Egypt itself moved the production of its Tucano to Helwan, deactivating the assembly line at Kader Factory.

All EMB-312s produced in Egypt received an anti-collision light on top of the tail. '518' is seen here at Helwan prior to delivery to the Egyptian Air Force. (Archive João Paulo Moralez)

A full flight line at Bilbeis, Egypt, with pilots grouped for a picture. (Archive João Paulo Moralez)

On 29 December 1988, the Government of Egypt signed the contract for the purchase of another 14 Tucano aircraft kits to be assembled by AOI. With this new sale, total aircraft for the Egyptian Air Force reached 54 units and the total of aircraft assembled in Egypt reached 124 airframes. The first kits of this order began to be delivered in May 1989 with the last kits arriving in December of that year.

Mixture of cultures

Culturally speaking, Embraer was not ready to receive the Egyptians in Brazil. The cultures of both nations were very distant when the Brazilian manufacturer received Egyptian technicians, pilots, mechanics and engineers at its headquarters, to provide them specialized instruction. Ary Guilherme Leber recalls those moments.

'The arrival of the Egyptians was impressive. We had never received people with such a different culture. We had to prepare a room where everyone stayed, each with a table. At the time they were here Ramadan happened, that is, they went to the factory and rested all day. After sundown they ate and drank. They came to the factory at Embraer, but slept on the tables. They did nothing else.

They also brought the rugs to pray at specific times. For that they had to wash their hands, feet and face. They did it in a sink, in a great juggling act to get their feet in the sink.

Usually they were all slow, they walked in a shuffled way and had a slower tempo. They were all good people, but had their own rhythm.'

In operation

Just as in Iraq, Egypt employed its Tucano fleet in basic training missions at the 101st Air Brigade at Bilbeis Air Force Base, 50km (27nm) away from Cairo. At that location, headquarters of the Egyptian Air Force Academy, students completed the primary phase in the HA Gomhouria (Bücker 181 variant built under licence in Egypt) and then moved on to basic instruction on EMB-312 Tucano. After 2000 the Tucano fleet was moved to 201st Air Brigade in Bilbeis. Embraer's aircraft received registrations from 501 to 554 and initially operated with an instructional colour scheme very similar to

In early 2000, the Tucanos received this interesting new overall white scheme with dark blue bands and details. (Archive Pit Weinert)

that of the Brazilian Air Force Academy, that is, white with an orange stripe on the fuselage. Subsequently, in the 2000s, the planes received a new overall white scheme with dark blue bands and details.

It is estimated that two airplanes have been lost in accidents with total loss. One during a taxiway exit, the second during a test with external loads. Those were Soviet weapons, in order to test the aircraft for use of armaments from that country. There was structural failure in flight as a result of 'flutter' killing the two crew members.

The Tucano continues to exercise the basic training task to this day, but was withdrawn from the intermediate stage with the arrival of the Chinese Hongdu K-8 training jets.

An interesting fact is that the Brazilian Air Force received two Egyptian instructors on 20 September 1989 as part of an exchange program between these two nations. On that occasion Colonel Mohamed Reda El Deeb Ahmed and Major Omar Mohamed Hassanein Naeim from the EAF came.

ARMÉE DE L'AIR

FRANCE

Armée de l'Air (AdA)

The history of the operation of the French EMB-312Fs in advanced training missions is closely linked to a purchase of French military helicopters by the Brazilian Army.

In 1987, the Brazilian Army began a study and technical analysis with the objective of acquiring two types of helicopters, utility and those for training and armed reconnaissance. The Brazilian Army had not had its own air assets since 1941, when its aviation was dissolved and merged with the Brazilian Naval Aviation to form the Brazilian Air Force.

Thus, the acquisition provided for a millionaire contract as the Army would still need spare parts, dedicated tooling and specific training for its pilots and mechanics to fly the new helicopters.

Finally, on 24 June 1988 a contract estimated at approximately USD 240 million involving 36 Aerospatiale AS.355K Panther and 16 Helibras HB-350L1 Esquilo paved the way for France to purchase, a few years later, 50 EMB-312F Tucano for the Armée de l'Air (AdA, French Air Force).

At the same time as the Army was studying the purchase of helicopters in mid-1986, France was seeking a replacement in the market for its domestically-manufactured Fouga Magister advanced jet trainers. In 1987, the tests and trials began with the main competitors and with those that would fit the requirements of France. Embraer, mindful of these needs that could produce a good contract beyond the prestige of having its trainer operated by a NATO-member country, held a demonstration for the Armée de l'Air along with sending a commercial proposal.

The three favourites at the time were, in addition to the Tucano, Pilatus with the PC-9 and Aerospatiale with a single-engine Omega turboprop developed from its Epsilon trainer, which was already in service with the Armée de l'Air.

The farewell paint scheme as applied to EMB-312F 459 '312-JD' featured artwork of a pilot on the tail bidding 'adieu'. (Chris Sacco)

A two-ship formation from the École de Pilotage et de Navigation (EPNA) 05.312, École de l'Air.
(French Ministry of Defence)

The unique characteristics of Tucano alone would not be definitive for France to make the choice in favour of the Embraer turboprop. Purchasing military equipment almost always depends on political and economic decisions. In this equation it would be necessary to add one more element for Embraer to win. And Switzerland helped Embraer.

At the time the Swiss government was looking for an alternative to its advanced training aviation and was evaluating the French Dassault Alpha Jet and the British BAe Hawk. The agreement, of course, involved trade-offs. If Switzerland opted for the Alpha Jet, France's purchase of PC-9 would be virtually automatic, even with Brazil having bought French helicopters for the Army. Trade compensations with Brazil would follow through other areas. But that was not what happened.

Switzerland opted for the Hawk and virtually made it unfeasible to sell PC-9 to France. Thus, the Tucano was the natural winner of the competition. At that time, EMB-321 Xingu was already in use, and it is still in operation today at French air arms.

Changes in the project

Although the Tucano was already a mature and proven product, France ended up requesting changes in the design to better tailor the operation of the aircraft in the country.

Initially speaking the fatigue lifespan would have to be increased from 6,000 to 10,000 flight hours by placing structural reinforcements on the fuselage and especially on the wings. Taking advantage of the refinements made on the Short Tucano wings, the French version also had removable and bird-impact resistant leading edges, as well as moving the refueling nozzle to a place near the leading edge to facilitate operation on soil without the need of ladders. The G load factor limit was extended to +7/-3.5G.

The avionics became of French origin; there was a significant improvement in the air conditioning system; the installation of a ventral aerodynamic brake and a propeller and canopy de-icing system.

Embraer's final proposal, including all modifications, was submitted in 1989 and the technical specification of the EMB-312F – as this version was renamed – was described in document ST-312F (F)/1015 issued by Embraer in January 1990.

Several meetings were held and the last one, where the draft contract was presented, took place at the Embraer subsidiary in France, EMBRAER Aviation International. The French government accepted the program in July 1990 when it entered official documentation and provision of funds. On 4 October 1991 the contract for 50 airframes with purchase options for another 30 was finally signed.

The planes would be delivered in five different batches. Only the first was confirmed and the others were to be confirmed between 1992 and 1995.

The first comprised two airframes for testing, a mock-up to integrate French avionics, spare parts, dedicated technical assistance, required technical documentation, four pylons and four drop tanks.

The second batch comprised 20 airplanes, plus four pylons and four drop tanks and technical assistance. The third, fourth and fifth batches would have 28, 15 and 15 aircraft, respectively.

Leading the issue of French avionics was Télécommunications Electroniques et Maritimes which co-ordinated Thomsom CSF/CNI, SFIM, Sextant, Rockwell-Collins France and other French companies.

Embraer made structural modifications to increase lifespan and G load factors in the airframe with c/n 312437, which was used for static tests only. The airplanes of c/n 312438 and 312439 comprised the two airframes of batch 1, pre-series and evaluation. The first one received the test registration PP-ZVC (later the French registration 438) and the second the registration PP-ZVD (439). On 7 April 1993, PP-ZVD flew for the first time and the PP-ZVC followed the next day. On 23 April, they were officially presented at Embraer, in São José dos Campos, before starting a series of test flights.

On 31 May, the planes followed to take part in Le Bourget. At the end of the fair '438' received the military registration 330-DJ, while '439' received 330-DM. Flight tests and technical-operational evaluations were conducted by the Centre d'Essais en Vol (CEV, In-Flight Tests Centre) at Brétigny-sur-Orge airbase. Weeks later, the planes were delivered to the Centre d'Expériences Aériennes Militaires (CEAM, Military Air Experiences Centre) at Mont-de-Marsan Air Base, at the Escadron d'Expérimentation et Transport 06.330 'Albret' (Experimental and Transport Squadron). There, France began to develop the operational doctrine of the advanced training of its future military pilots. This process took a year and by that time France had already requested batch 2, with 20 airplanes. The EMB-312F 438 followed to École de Pilotage et de Navigation de l'Armée de l'Air 05.312 (EPNA 05.312, Pilot and Navigation School) at the Salon de Provence Air Base, around 56 km from Marseille, on 4 July 1994.

Delivery flights

Delivery of the planes was made in flight, crossing the Atlantic and with the EMB-312F leaving São José dos Campos to Paris. At the French capital, at the headquarters of EIA, airplanes would be prepared and officially delivered to the Armée de l'Air.

One of the most striking Armée de l'Air paint schemes was this dark grey adopted to make the Tucano more conspicuous to other aircraft at long distances. (French Ministry of Defence)

The 23 October 1994 was marked at Embraer by the take-off of the first three airplanes for the long-range flight. The EMB-120 Brasilia PT-ZBA would continue to lead the way by supporting the planes and their crews and also taking spare parts and materials needed for this type of event. The route would include São José dos Campos, Recife, Fernando de Noronha, Sal Island, Gran Canaria, Seville and Paris.

For these flights the airplanes were equipped with radios VHF and GPS, among others devices. Drop tanks held 400 litre (105gal) instead of the usual 330-litre (87gal) used by Embraer aircraft. These new tanks were made by adding two sections to the 330-liter (87gal) tanks of the Xavante AT-26, to increase its length and internal fuel capacity. Usually only three planes followed per trip.

The first production and series airframes were delivered on 18 January 1995 when they were assigned to Division des Vols 05.312 da École de l´Air. This division was after designated École de Pilotage et de Navigation (EPNA) 05.312 from 2004 to 2008, changing to Centre de Formation Aerienne Militaire Initial (Initial Military Air Training Centre - CFAMI) after that. The Groupement École GE315 (School Group) at Cognac Airfield 709, where the instructors were trained, also received the Tucano.

Following this, France confirmed the second batch of 28 aircraft, but did not exercise its 30 options through batches 4 and 5. Adding the two aircraft of the first batch the country received 50 EMB-312F. It took two years of deliveries, with the last planes leaving Embraer on 27 November and arriving in Paris on 3 December 1996.

In operation

In the 1990s the training of cadets in France was divided into several stages. First the student would make primary and basic flights in the CAP-10 at the Air Base 701 of Salon de Provence, following to Cognac where they received the advanced instruction in the Aerospatiale Epsilon. When declared aspirants, the students then divided according to the aviation branch they would follow from then on. For those who would become future fighter pilots instruction began with the Tucano in a 60-hour program followed by Alpha Jet on advanced flights at the École de Chasse at Tours Air Base 705.

The EMB-312F was based in Salon de Provence, on the 2e EIV 'Montmirail' and 3e EIV 'Côte Bleue' of Division des Vols 05.312, named after École de Pilotage et de Navigation EPNA 05.312 and the Centre de Formation Aerienne Militaire Initial of École de l'Air. In that place, replacing the old Fouga Magister, the pilots did aerobatics, training flights, VFR and IFR navigation. In addition they had navigation on long-range and night flights. In total, it was a 15-week training course on the Tucano at Salon de Provence with some 9,000 hours annually. From those 15 weeks, three was dedicated to ground school, plus nine flights for the student to fammiliarize with the plane. Plus eight aerobatic flights, two of them solo, followed by more flights in two ship formation, eight with instructors and four solo.

After that there was another 20 flights on the simulator for normal and emergency cabin procedures. Then, 21 flights on VFR, IFR, low-level flying and ground controlled navigation in 45 meters (150 feet) at 166km/h (90knots). There is also the night navigation phase, first with an instructor and the last four were solo. Total was around 50 flights in 60 hours.

Pilots of French Navy also flew the Tucano in first stage of the fighter pilot course and the École received, annually, around 100 pilots in five or six class. Avarage is 250 flight hours per airplane per year. Two EMB-312F were lost in accidents during its operation.

Initially the colour used in French instruction was white and red, however studies pointed out that the colours did not make it easy to see the planes in flight over great distances. Thus, a dark grey was adopted later.

After that, the Tucano was prematurely retired when the fleet still had two-thirds of its flight hours still available. This was mainly due to the constant reduction of funds from the French military budget and the reduction in the demand of future military pilots. Thus, the Armée de l'Air underwent a profound restructuring including the outsourcing of various services in the air force.

However, it is noteworthy to say that the Tucano was not well regarded by several ranks in the Armée de l'Air, who saw their purchase as the death of the Aérospatiale Omega. With the departure of Tucano in July 2009, Epsilon took over its role.

Currently the Armée de l'Air cadet receives the primary and basic training in the Cirrus SR-20/22, moving on to the Grob G120A and then to the Epsilon in the first stage of fighter pilot training.

Meanwhile, France is gradually selling its planes to other operators. In addition to donating six examples to Mauritania, 13 were sold and registered in the US with civilian operators, while others have been preserved at museums.

HONDURAS

Fuerza Aérea Hondureña (FAH)

The first nation in the world to acquire the EMB-312 Tucano, after Brazil, was Honduras.

The interest of the small Caribbean country by the Embraer trainer arose during a technical stop that one of the EMB-312 demonstrators, PP-ZTC, made on Honduras when it went on a world-wide exhibition tour in March 1983. The purchase was made yet in that year and before 1984, Honduran pilots and mechanics were already taking the course on the aircraft at Embraer, in São José dos Campos.

For political reasons, the contract was not announced. On the contrary, everything was kept in absolute secrecy.

At that particular moment the governments of Honduras and Nicaragua were in full tension. Supported by Cuba, the Sandinista movement that was in power in Nicaragua was waging war against groups of guerrillas, who were trying to overthrow the government of that country. The Contras, as these rebels were known, received strong US support that indirectly provided weapons and equipment so that the struggle continued until the Sandinistas' fall was achieved.

The US Congress forbade the country to support the Contras. But if officially this support did not exist, US officials secretly allowed the rebels to be fueled through another path. It was through the Iranian government.

Almost all of Iran's military equipment was American-made. After the Islamic Revolution broke out in 1979, the United States enforced several embargoes that greatly affected the Iranian military and warfare capacity. With a total ongoing war against Iraq, Iran needed at all costs to receive new stocks of armaments and spare parts to keep its military arsenal running. Thus, the US facilitated this supply with Iran by funding the Contra in Nicaragua. The operation was exposed in the mid-1980s and became one of the biggest US scandals in the 20th century.

In this turbulent scenario Honduras, which received much economic support from the US, allowed its territory to be used by the Contra through bases of operation. Hon-

For interception missions, the Honduran EMB-312 fleet uses the FN Herstal Twin MAG pod with two MAG 58P 7.62mm machine guns and 1,000 rounds per pod. It is common for the Tucanos to operate as detachments from different airfields. (Oscar Elvir)

More recently Honduras adopted this dark upper surface scheme on some of its EMB-312s. FAH 251 carries two seven-tube 70mm (2.75in) rocket pods. (Oscar Elvir)

duras also faced its own internal instability with the presence of left-wing groups that attempted to destabilize the Honduran government and still watched with concern the civil war that was in full swing in El Salvador, another neighbouring country.

Brazil has always adopted a neutral policy without supporting any political or ideological movement, much less furnishing military equipment to further foment tension or conflict.

Thus, the EMB-312 sale should be kept confidential. And everyone was able to do it very well. No news on the subject was leaked to the press. But on the delivery flight of the first batch of aircraft – four in total and which were backed by a Lockheed Electra II from the Fuerza Aérea Hondureña (FAH, Honduaran Air Force) itself – the case came to light in May 1984. This squadron of Honduran military aircraft was caught during the technical stop at Boa Vista, still in Brazilian territory.

The repercussion was immediate and demanded a position from the Brazilian authorities who claimed it to be merely a training aircraft sale contract, without the supply of weapons. Because it was not dealing with war material itself, it did not violate the criteria of the National Program for the Exportation of War Material. In July the final four aircraft were delivered in the second batch also transferred in flight. In 1985, four more EMB-312s were ordered by Honduras – bringing their fleet to 12 aircraft. This last order was delivered in July 1986.

Upon arrival in Honduras, the EMB-312 Tucano replaced the North American T-28 Trojan. These planes were very old-fashioned, difficult to operate because there were no spare parts, they were expensive to maintain and even to fly.

The aircraft were assigned to the Escuela Militar de Aviación (Military Aviation School), later Academia Militar de Aviación (Military Aviation Academy), at Colonel José Enrique Soto Cano Air Base and at 1° Escuadrón de Entrenamiento, located in Palmerola, Comayagua and sporting a paint scheme in shades of Marine Green, green, rusty red and field drab. The planes were registered from FAH 250 to 261.

Upon entering the academy the Honduran cadets initially flew the Cessna 172H (T-41). Advanced training and operational conversion for jets and combat aviation, including the Cessna A-37 Dragonfly and Northrop F-5E/F were made in the Tucano. In these cases the trainings included tactical navigation, day and night flights in forma-

Honduras maintained a partnership with Colombia to aid interception of illegal or irregular flights. Here, a FAC Cessna Citation SR-560 Tracker, used for intelligence tasks and AEW&C, flies with Honduran EMB-312 FAH 254 during joint operations. (Oscar Elvir)

tion, IFR flight with weapons, interception and ground attack doctrines, totaling 150 hours.

But the operational flexibility also allowed the Tucano to be used on patrol, reconnaissance, escort, search-and-rescue (SAR) for aircraft lost at sea or in the jungle, interception and air defence.

Since 2000, Honduras has enforced and intensified the interception of small aircraft flying without flight plans and entering its airspace without authorization. Keeping aircraft alert and armed, one of the first occasions when it succeeded in this type of flight was the interception and downing of an Aerocommander 500 on 14 April 2003. The air traffic control of Honduras called the aircraft's crew on the radio, but they did not respond to the calls. A FAH EMB-312 was sent to recognize registration and nationality and to force the Aerocommander to land immediately for investigation. As there was no response, the Tucano opened fire and shot the twin engined aircraft down.

In January of the following year a single engined aircraft was also shot down by a FAH Tucano. The aircraft remained without a flight plan and also did not respond to calls from the country's air traffic control. FAH pilots, while intercepting the aircraft, reported that the occupants on board dropped the cargo in mid-flight. The aircraft fell in flames on the town of Agua Zarca, in Valladolid, Lempira.

Little is known about how many planes Honduras has shot down over the years. This information is classified, however reports indicate that the country's EMB-312 alone were responsible for shooting down 40 aircraft in the service of drug and weapons dealers. Basically the armaments used for both operational training and routine operational employment are summarized as FN Herstal's Twin MAG Pod with two 7.62mm (.30in) MAG 58P machine guns each. In addition, 70mm (2.75in) rocket launchers with seven units per pod are used.

Throughout its operation only three airplanes were lost in accidents. The first, FAH 261, crashed on 9 December 1987, with loss of its pilot in the town of La Guama. The second one, FAH 255, crashed on 5 October 2000 in the city of Colón, killing both crew members.

Finally, Tucano FAH 253 was lost on 3 November 2004 in the village of San Jacinto de Carbonales, in Comayagua, again with the loss of the crew.

IRAN

Islamic Revolutionary Guard of Corps Air Force, (IRGC AF, 1979-2011)

Islamic Revolutionary Guard of Corps Air & Space Force, (IRGC ASF, 2011 until today)

The elite Islamic Revolutionary Guard Corps (IRGC) force was established on 5 May 1979 to protect the country's politics, ideals, and political and organizational system from outside interference or even attempted coups d'état. The IRGC operates entirely independent of the regular armed forces of Iran. It has its own equipment, doctrine, bases, structure and model of action with three armed branches with the function of rapid mobilization and tactical support, being a naval force, an army and an air force.

Negotiations with Embraer and the Brazilian government to purchase the EMB-312 with the exclusive purpose of advanced training of IRGC AF pilots began in mid-1988 when the Iranian government was about to end the eight-year war with its neighbour Iraq. Later that year, an estimated USD 160-million contract was concluded with Iran for 15 EMB-312 airframes. The planes received the camouflage in shades of tan, green and brown at the top and light grey at the bottom. The aircraft received Iranian registrations 15-2401 to 15-2415.

A very rare picture showing Iranian EMB-312 15-2407 with an unguided rocket pod. (Ahmad Mahgoli, via Babak Taghvaee)

Since its introduction in Iran, the EMB-312 fleet has used the same desert camouflage.
(Archive Babak Taghvaee)

Training in Brazil

The training of 38 pilots and mechanics was given by FAB soldiers at the Lagoa Santa Aeronautical Material Park and was surrounded by interesting stories to tell.

'The Iranians wanted instruction from military pilots. It was a demand. And it had to be far away, in the case Lagoa Santa (MG). They stayed in the barracks at the base, but were very bad at flying. They had already done ground school on the Tucano and most had 500 hours of flying experience. Some had 1,000 hours and had experience with the Pilatus PC-7, which is similar to the Tucano.

But they showed up without their flight suits, they were all wearing the orange Embraer flight suit. Almost everyone wearing regular shoes and two came in flip-flops. We called Embraer and requested flying boots with steel reinforcement, because in the ejection they could have their foot pulled out.

That's when we started flying. But their basic flight techniques were very limited. Many did not know how to spin so we had to teach everything from there', recalls Major Vilarinho, who led the Tucano training for Iranians.

During one of the flights, an incident ended up marking the passage of the Iranians in Brazil. During a slow tonneau manoeuvre, when the Tucano was flying upside down, one of the students ejected. The action was involuntary with the handle being accidentally pulled by the student while crew and airplane were suffering negative G.

The aircraft suffered little damage and returned to Lagoa Santa with only the instructor in the rear seat, but with the canopy broken, without a seat and fear stamped on the face of the Brazilian instructor. The Iranian pilot landed 120km (64nm) away from the base suffering minor injuries. He was rescued by three farmers and it was enough to pronounce 'Lagoa Santa' to be sent back to the base.

The training was all done on Iran's own aircraft and they were already painted in operational colours, but without military insignia and registration.

Thus, even though they were far away and in a discreet military base, the planes were caught in operation in Brazil using provisional Brazilian civilian registrations. The photos were published in the country's largest national weekly magazine. So far the negotiations were kept secret since Iraq, Iran's enemy, also operated exactly the same plane.

The Iranians had the chance to visit Embraer headquarters in some situations. On one of these occasions, the trip to São José dos Campos combined exactly with the passage of an Iraqi delegation also at Embraer. On the same day, at the same time, military of the rival countries almost found themselves in the corridors of the Brazilian manufacturer. Fortunately the employees succeeded in separating the groups so that they would not meet. Everything worked and it would be impossible to predict the outcome if they crossed each other's path.

In operation in Iran

There is no evidence or documents that indicate how the EMB-312s were taken to Iran. The most likely is that they have been transported by plane.

In that country the aircraft were concentrated at the Badr Air Base in Isfahan, where the IRGC AF flight training centre is located. To date, only one plane has been lost in an accident during a flight in bad weather. It is believed that close to 10 aircraft are in flight conditions with four stored to preserve the age of the fleet.

In addition to the training and training of pilots, Tucano's operational flexibility also included attack missions, armed reconnaissance and CAS.

In the internal security section of the country, airplanes were used to fight drug dealing and also against Taliban fighters at various points along the border with Afghanistan, especially in the first half of the 1990s.

The easy maintenance and the possibility of being operated on rudimentary airstrips, land or sand and with little ground assistance made the country take advantage of these characteristics and simultaneously used the aircraft in secondment, with two or at most four aircraft at various locations such as Kashan, Orumiyeh, Shiraz and Zahedan.

The motives were almost always advanced and operational training with the use of weapons, however in almost all these occasions the presence of the airplanes was used to conduct reconnaissance and counter-terrorism flights.

The crews are trained for joint operations with IRGC AF's other aircraft and combat helicopters, such as Bell 214, Bell AH-1J, Mil Mi-17 and Sukhoi Su-25.

The armament basically consists of two launchers with seven 70mm (2.75in) rockets each, taking advantage of a doctrine absorbed from the USA through a military mission that occurred in the 1960s in that country. At the time, despite modern aircraft such as the F-84, F-86F and T-33A, Iran was unable to combat groups of Kurdish guerrillas hiding in the mountains in the region bordering Iraq that were successfully attacking the regular army and were fortifying more and more at that point, threatening an expansion of its area of operation.

A belly view of EMB-312 15-2413. In Iran, two 70mm (2.75in) rocket pods are only carried on operational missions. (Ali Reza Khodakarami)

The morale of the troops was very low, the willingness to fight even lower. Air force jets were not available to operate from that region, and there was little autonomy to actually stay over the targets. The solution was found in 15 T-6s that were armed with 7.62mm (0.3in) machine guns and 127mm (5in) Zuni rockets. Taking off always in pairs from the Kermanshah Airfield, the T-6 made punctual, precise attacks and before the Iranian troops, whose morale began to rise again. In a short time, with air support, the Kurds were defeated and the threat of the frontier was lifted.

The legacy of this North American mission remained and thus the Tucano also managed to successfully fight the guerrilla groups of the Kurdistan Workers' Party.

From 2008 to 2010 new engagements were made against the Islamic extremist group Jond-Allah in the town of Zahedan in joint operations with the Su-25. In the same locality the EMB-312 also made attacks later, against the Islamic extremist group Jaish Al-Adl.

It is rumored that Iraq sent a quantity of EMB-312 to Iran during the US invasion of the Gulf War. However recent Iraqi sources deny this fact by saying that no Tucano was actually taken to Iran.

IRAQ

Al Quwwat Al Jawwiya Al Iraqiya (IrAF, Iraqi Air Force)

Having been the largest operator of the EMB-312 in the world after Brazil itself, Iraq received in all 80 copies that operated with the registrations 6701 to 6780.

All Iraqi airframes were assembled by Egypt under a production licence provided by Embraer. In fact, the first EMB-312 assembled in Egypt, at Kader Factory, was an airframe for the Iraqi Air Force, aircraft c/n 312105 and it received a paint scheme extremely similar to that used by the Brazilian Air Force Academy, that is, white with orange bands.

That, in fact, was the first airplane in the history of Embraer produced outside its facilities in São José dos Campos. The maiden flight of this airframe was commanded by the Egyptian Air Force General and test pilot Gamal Farghaly, lasting 40 minutes on 8 July 1985. In the following months a further two units, which would also be destined to Iraq, were finished in Egypt and delivered to the Baghdad government on 11 October 1985.

But Iraq had already received 10 units of Tucano the previous year. Under the agreement signed between Cairo and Brazil, the first 10 aircraft of the under-licence production agreement would be assembled in Brazil with the assistance of Egyptians and flown to Iraq. These airframes had a paint scheme extremely similar to that used by the Smoke Squadron in Brazil, that is, all red with white bands and black outlines.

Arriving in Iraq the aircraft replaced the already obsolete Czech advanced jet trainer Aero L-29 Delfin. In that country the aerial training proceeded as follows. The primary training, around 8 hours, was in the AS-202/18A2 Bravo in the first year of instruction at the Iraqi Air Force Academy in Tikrit. In the second year the student had contact with the Tucano in the intermediate formation flying around 115 hours performing navigation, training, aerobatics, instrument and night stages. Then, in the third year, the student would go to the L-39 where he would fly another 70 hours in advanced instructions learning combat manoeuvres, interception, attack, weapons use and others.

Five Iraqi EMB-312s ready to be delivered at the Arab Organization for Industrialization (AOI) factory at Helwan, Egypt. (Archive Mário Vinagre)

Of the 80 aircraft received, only 40 were kept operational while the others were stored in fortified shelters to maintain the average age of the fleet and to replace planes lost in an accident or that had suffered a serious problem or were grounded to undergo a major overhaul that could take months to complete.

In Iraq, airplanes were used exclusively for training and never used weapons or were used for operational flights.

During the Second Gulf War (August 1990 and March 1991), many aircraft survived Coalition attacks authorized by the UN Security Council, but did not resist the 2003 US-led invasion. Many airplanes were destroyed and their operation in Iraq was impossible to resume. Around 12 EMB-312s were found at Al-Fat'h Air Base, while almost all the rest was at Tikrit.

An Iraqi EMB-312 in its training white and DayGlo scheme during a test flight with a still-incomplete registration. (Embraer)

KENYA

Kenya Air Force (KAF)

Short Tucano's first export customer was Kenya, which acquired 12 aircraft in 1988 to replace the onerous and already difficult to maintain BAC Strikemaster 167 Mk 87 in the advanced training and light attack missions.

Designated by the manufacturer as T.Mk 51, the aircraft would be used in the operational conversion of the newly formed pilots in the Scottish Aviation Bulldog primary and basic training aircraft to the country's combat jets and air defence fighters in a program similar to RAF training. In this case, the pilots would learn all the techniques of interception, tactical navigation, combat manoeuvres, air-to-air and air-to-ground doctrines besides making use of real weaponry.

Kenya's first production model flew for the first time on 11 October 1989 with the provisional registration of ZH203.

Military registrations at KAF were from 810 to 821, later changing from 811 to 822. This change happened because one of the conditions for the acquisition of these airframes by the air force was that they should be made compatible with the use of armaments on four underwing hard points. This way, Short Brothers initiated a wing modification program and a Tucano weapons certification campaign including the 12.7mm (.50in) machine gun pods, general purpose bombs of the Mk 81 and Mk 82 family and 70mm (2.75in) unguided rocket pods with seven units each.

In COIN configuration the plane had an operational radius of 732km (390nm) allowing 5 minutes at maximum power over target.

On 22 February 1990, the campaign costed the life of Alan Deacon, Short's chief test pilot, who crashed into the sea when he was piloting ZH203 on a test flight carrying cargo under the wings. As the aircraft had not been delivered to Short Brothers it produced a new aircraft and KAF chose to skip the number 810, starting from 811.

Deliveries began in June 1990 and half of the fleet received a typical painting very similar to RAF's first standard of instruction – red and white, while the others received a tactical camouflage, in grey and green, following the air force's combat paint scheme. All the aircraft were allotted in the Kenya Air Force Flying Training School (KAF-FTS),

A flight of four Short Tucanos in grey and green camouflage at Exeter Airport prior to their delivery flight. (Via Georg Mader)

A three-ship formation of Kenyan Tucanons during an air parade. (Archive João Paulo Moralez)

which was established in 1940 by RAF at the former Eastleight Air Base – now Moi Air Base, at the Mombasa International Airport, in the suburbs of Nairobi.

Although they were also purchased with the purpose of executing counterinsurgency (COIN) missions, the Short Tucano never entered combat and were rarely moved to patrol flights in the country's border region or areas of national interest. After 2010, KAF purchased three sets of smoke generator system, that were installed one under each wing of the aircraft in order to create greater visual effect during the air presentations on national holidays.

KUWAIT

Al Quwwat al Jawwiya al Kuwaitiya

After Kenya, Kuwait was the second and last export customer to purchase Short Tucano for the primary and basic training of its future combat pilots including tactical flights and training for weapons use. The turboprops were selected to replace the already obsolete BAC Strikemaster within the Kuwaiti Air Force (KAF).

In February 1989, the purchase of 16 aircraft was made. On 21 September 1990, the first production aircraft with provisional registration ZH506, had its maiden flight.

The Kuwaiti variant was designated T.Mk 52 by Short Brothers and received a desert camouflage painting (sand and brown) with the leading edges of the tailfin, wing tips, horizontal stabilizer tips and propeller spinner in orange.

Deliveries were scheduled to take place in 1991, but were delayed due to Iraq's invasion of Kuwait and the ensuing Second Gulf War. The airplanes were moved from Belfast to Shawbury, where they were in stock and preserved condition until a decision was made as to their destination.

Anyway, in 1994 the order was given for delivery to happen at the earliest possible time. The aircraft underwent an overhaul, were transported back to Belfast and eventually delivered in two batches to Kuwait in April and May 1995.

All airframes were assigned to the 19th Flight Training Squadron at Ali Al Salem Air Base, in the country's central region. They were allocated serial numbers from KAF 101 to 116. The aircraft were never used for combat missions. However, for some 20 years they operated with a wide range of weapons cleared for the Short Tucano, to qualify future military pilots. Today, all Kuwaiti Tucanos are stored. It is unclear if they ever return to operational status.

Tucano KAF 108 initiates a break during a training sortie. No national insignia or registration appear under the wings. (Archive João Paulo Moralez)

A rare close-up of Kuwait Air Force Tucano T.Mk 52 KAF 112.

MAURITANIA

Force Aérienne Islamique de Mauritanie (FAIM, Air Force of Mauretania)

With the deactivation of the Armée de l'Air EMB-312Fs in 2009, more than 36 aircraft were stored in France awaiting a final destination.

Because they still had a considerable amount of remaining flight hours, about two-thirds of their available life, since the Tucano was prematurely deactivated in France, the government of Paris decided to donate five airframes to the Force Aérienne Islamique de Mauritanie (FAIM, Air Force of the Islamic Republic of Mauritania). The aircraft were prepared to perform advanced weapons training and mainly to conduct patrol flights in the border region, intercept light aircraft violating airspace and attack positions of guerrillas and insurgents, mainly from al Qaeda in the Islamic Maghreb (AQIM) who was mainly engaged in kidnapping, localized attacks and drug and weapons dealing.

France donated the planes after making a thorough overhaul and changing the paint scheme to a desert camouflage in two shades of sand. FAIM selected a group of pilots that was sent to France to train in the EMB-312F, with deliveries made in 2010 and 2011.

The aircraft were transferred on mixed crew flights (French in the rear seat and Mauritanian in the front) with a stop-over at Torrejon de Ardoz Air Base in Spain, before continuing straight to the Base Aérienne 210 Atar (210 Air Base) in Atar in the middle of the country and being destined for the École Militaire de l'Air (Military Air Academy).

They were hardly received and the aircraft already played a significant role in fighting terrorists by attacking a column of 20 AQIM vehicles in September in the desert region of the country near Hassi Sidi, near the border with Mali. Two terrorists of the AQIM high command were captured while 12 combatants were killed.

Already in March 2011 the EMB-312F took action, when they carried out an attack on a column of AQIM vehicles in Mali. The countries have a cooperation agreement to combat terrorists that allow one country to enter in the territory or airspace if its forces are pursuing terrorists.

Also in October that year, again in Mali, one of the AQIM leaders, Ould Sidi Aly was killed in an air raid by the Mauritanian Tucano. The terrorist was the mastermind of AQIM who planned the attacks and was on the list of most wanted by the countries since 2007. The actions took place in the Wagadou forest, which covers Mali and Mauritania.

The FAIM lost two Tucano in accidents. The first was on 21 August 2011 in the town of Chinguetti with the loss of its occupants. France sent a new airframe to replace it; however on 18 March 2013 another plane was lost killing the mechanic who was in the aircraft in a maintenance flight. The EMB-312F crashed 7km south of Aoujeft in the Adrar region.

EMB-312F 5T-MAI during a training flight over Base Aérienne 210 Atar. The Tucanos were often used to strafe rebels, armed with two FN Herstal Heavy Machine Pods with M3P 12.7mm (0.5in) machine guns and 250 rounds. (Archive Pit Weinert)

Today, three airplanes are in operation and one was scraped to provide as spare parts for the fleet. With the arrival of Super Tucano in 2012, the EMB-312F in Mauritania was concentrated in advanced training only.

Altought the EMB-312F has two hard points, the original provision for Armée de l'Air was only about fuel tanks. A modification was necessary to install electric wire and a armament panel at the cockpit. Mauritania operates only with 12.7mm (.50) machine guns, two of them per aircraft.

PARAGUAY

Fuerza Aérea Paraguaya (FAP)

The EMB-312's multi-role capability attracted the attention of the Paraguayan authorities, who opted to purchase six Embraer turboprop aircraft in 1987 with weapons-carrying capability and a Vietnam-style camouflage.

Because Embraer was in full production rate of the EMB-312, the planes were delivered that same year to Paraguay with the registrations FAP 1051 to FAP 1056. The transfer was done by the Paraguayan and Embraer crew and with a technical stop in Foz do Iguaçu on 29 December 1987.

With the arrival of the new airplanes to the Fuerza Aérea Paraguaya (FAP, Paraguayan Air Force) the country created the Tercero Escuadrón de Reconocimiento y Ataque 'Moros' (3rd Reconnaissance and Attack Squadron 'Moros') that had under its organization the squadrons Gamma and Omega. The new squadron was subordinated to the FAP Grupo Aerotáctico (GAT, Aerotactical Group) and based at Ñu Guazú Air Base in Luque, near the capital Asunción.

Tucano's missions in that country were advanced training for the instruction of fighter pilots, Paraguayan airspace patrol, attack, interception, escort and armed reconnaissance. Because it also carries out operational and combat missions, Tucano was assigned the AT-27 designation in that country, the same used by Brazil in the 2ª ELO.

In 1988, the Tucano flew almost 1,035 hours, and the following year, when almost 1,247 hours were flown, FAP lost its first EMB-312 in an accident on 19 October 1989. It was a navigation instruction flight in circuit in the cities of Asunción-Ciudad del Este-Saltos del Guairá-Curuguaty-Asunción. The EMB-312 FAP 1056, under the command of Lieutenants Juan Antonio Rojas Duré and Blas Rodríguez Rivas, crashed at the airport near Itaipu bi-national hydroelectric plant culminating in the death of both crew members.

The weapons load of EMB-312 FAP 1052 includes rocket pods, machine gun pods, 118kg (400lb) incendiary bombs, 330-litre (87-gal) fuel tanks and BEX-11 exercise bombs. (Antônio Luís Sapienza Fracchia)

EMB-312 FAP 1053 displays its BEX-11 exercise bombs soon after take-off from Ñu Guazú Air Base in Luque. (Antônio Luis Sapienza Fracchia)

In 1991, that unit happened to be designated Tercero Escuadrón de Caza 'Moros' (3rd Fighter Squadron 'Moros') maintaining its original flights. That was the year that most EMB-312s flew throughout their career at FAP. There were almost 1,457 hours in all. The following year a new restructuring in FAP subordinated the GAT to the 1ª Brigada Aérea. The precarious economic situation of the country affected the operationality of the planes, which in 1994 accumulated almost 522 hours of flight. But even with budgetary restrictions and flight hours, the 1052, 1053 and 1054 aircraft were deployed and operated in a secondment basis from the city of Ayolas for reconnaissance and training flights in that region. On 5 March, however, FAP 1054 manned by Captain Williuam Sánchez and Mr. Edgar Cramer, son of the then FAP commander, crashed near Ayolas Airport, leading to the death of the two occupants.

Later that year, '1051', '1052' and '1053' flew to Argentina to replace the canopy with pieces made in Argentina. It was of poor optical quality causing distortions when looking through it, having in addition a different format that led to greater aerodynamic drag.

After the loss of the two aircraft in an accident in 1995, FAP commissioned another six new EMB-312 from Embraer, but withdrew from the purchase when the planes were ready, painted and registered in São José dos Campos awaiting delivery. These aircraft were stored and then sold to Angola in 1998.

On 10 March 1998, FAP would still resent another fatal accident that killed Lieutenant Néstor Sánchez and maintenance sergeant Mario González, who were aboard 'FAP 1055' in the town of Ayolas. As a result, the country had only half the amount of airplanes it had acquired in 1987. In 1999, Brazil again supported Paraguay by overhauling and maintaining the remaining three Tucano – FAP 1051, 1052 and 1053 – when it received the planes at Lagoa Santa Aeronautical Material Park. The aircraft remained for about a month and then returned to Paraguay to participate in training missions and operational flights from Concepción Air Base and Mariscal Estigarribia Air Base, a strategic location in the Chaco region.

EMB-312 FAP 1052 approaches the test range. It is fully armed in close air support configuration with FN Herstal Twin MAG Pods and 70mm (2.75in) rocket pods. (Antônio Luis Sapienza Fracchia)

In 2002, there was the largest Tucano operational decrease in Paraguay, as only 27 hours been flown. The squadron got back on its feet and in the next year it improved its operability again.

Aiming to develop a joint doctrine for the efficient coordination and interception of small aircraft on illegal flights serving the smuggling of merchandise, drugs and weapons, Brazil and Paraguay held the binational exercise PARBRA (Paraguay-Brazil) from 10 to 14 June 2005. Involving the two air forces, Brazil participated with A-27s from 3°/3° GAV Arrow Squadron of Campo Grande Air Base, while the Paraguayan AT-27 departed from Concepción Air Base. A Paraguayan military transport aircraft took off and entered Brazilian airspace without permission and flight plan. After being detected by Brazilian radar, the A-27 were sent to intercept it and forcing it to land in Campo Grande for investigation. In the same way, Paraguay, using their Tucanos, did the same procedures, forcing a FAB transport plane to land at Concepción.

Thus, both countries could establish a series of standard procedures during interceptions.

This exercise was repeated from 18 to 22 May 2009 and then again in 2013, when Brazil participated using the EMB-314/A-29 Super Tucano.

From 19 to 23 December 2012, Paraguay held the binational exercise ARPA I (Argentina-Paraguay) with Argentina, in the same model and similar to PARBRA.

In 2010, Paraguay received three fromer FAB EMB-312s in exchange for a Boeing 707 and four Xavante AT-26 cells. It took months of talks and negotiations until the authorities of both countries could reach a final agreement.

Thus, on 10 November, EMB-312 with registration number FAP 1057 (ex-FAB 1319) was delivered to Paraguay and on 29 December, FAP 1058 and 1059 (ex-FAB 1344 and 1347) delivered by Brazilian pilots followed.

It is worth noting that ex-FAB 1319 and 1344 had been lent to Argentina for a period of one year, from 2009 to 2010. Thus, these airframes served in three different air forces during their operational careers!

FAP 1053 during a training mission. The low-quality Argentine-made canopy can be clearly disdinguished. (Luciano Porto, www.spotter.com.br)

On that occasion the then president of Paraguay, Fernando Lugo, with FAB's Chief of Staff Brigadier João Manoel Sandim de Rezende representing Brazil, participated in an official delivery ceremony.

Throughout its operation Paraguay has used a formidable range of weapons for its aircraft including 70mm Avibras 70/7M5A or Equipaer EQ-LMF-70/7AP rocket launchers with seven tubes each; Mk 81 and Mk 82 bombs; FN Herstal Twin MAG Pods, Avibras M2A1 180kg (400lb) napalm bombs and 7.62mm Twin MAG Pod FN Herstal machine gun pods. Mk 76 Mod.5 of 11kg (24lb) practice bombs with smoke cartridges are used in instruction flights.

PERU

Fuerza Aérea del Peru (FAPe)

Alongside Honduras, Peru was the nation that most intercepted and shot down light airplanes used on illegal drug flights. There were more than 90 interceptions and around 40 planes shot down using 12.7mm (.50in) machine guns. Another five were forced to land for inquiries. Some 40 clandestine airstrips used for this type of flights in remote locations or in the Amazon jungle were also destroyed.

But the arrival of the EMB-312 in Peru did not initially have this objective.

In 1986, the Andean country acquired 20 EMB-312 airframes to conduct operational instruction for fighter, attack and transport aviation. The contract value was estimated at USD 32 million including spare parts, pilot and mechanic training and maintenance tooling.

The first four airplanes were ferried in April 1987 and the last in November of that year. All of them had a training paint scheme in white and orange and were allocated in the Escuadrón Aéreo No. 512 (No. 512 Squadron) del Grupo Aéreo No. 51 (No. 51 Air Group), in the Las Palmas Air Base. Four years later the planes were transferred to Pisco Air Base, 600km south of Lima, the capital of Peru.

In Peru students entered the Escuela de Oficiales de la Fuerza Aérea del Peru (Peruvian Air Force School) at Lima, where they did all the theoretical part and primary flights in the Cessa T-41 (Cessna 172) with 30 hours until reaching the solo flight. Upon being a declared officer candiadate (ensign), the students went to the Grupo Aéreo 51 where they did the basic and intermediate stage in the Tucano comprising 150 hours including navigation, IFR, night, training and aerobatic flights.

An EMB-312 flight line at Pisco. All aircraft are in white and DayGlo training colours. (Lewis Mejia Prada)

Upon completing this stage the officer candiadates received a more advanced and specific instruction in other models according to the function they would perform in Fuerza Aérea del Peru (FAPe, Peruvian Air Force). For the fighter and attack pilot the continuation of this training followed in the Aermacchi MB-339A, for example.

In 1998, the Zlin went into operation in Peru modifying the training system a little. After the T-41 at Escuela de Oficiales de la Fuerza Aérea del Peru, cadets flew in Grupo 51 Zlin Z-242s on basic flights (50 hours), in Escuadrón Aéreo No. 511, and intermediate in Tucano (100 hours), in the Escuadrón Aéreo No. 512. They then proceeded to the MB-339A if they were assigned to fly fighters or attack aircraft.

Closing the siege against drug traffickers

But the arrival of the 1990s caused the Peruvian government, especially motivated by then President Alberto Fujimori and yielding to US pressures, to initiate a more rigid policy on flights of airplanes carrying illegal weapons and drugs. Thus the FAPe was prepared to restrain these types of flights with more vigor. Another 10 planes would be bought just to fulfill this mission and Tucano clearly was the favourite at that point. In order for the government to authorize this purchase, it would be necessary for the Tucano to be approved by an operational technical evaluation carried out by FAPe.

So, the FAPe agreed to lend airframes FAP 423, 424 and 473 to the mission. Embraer sent a technical team that made the modifications in the airplanes installing wiring in the wings, sight and pylons to carry weapons. In addition, the FAP 429 was used as a chase airplane.

From 28 to 31 August 1989, the modified aircraft began the tests that totaled 48 hours of flight and 33 sorties with mixed crews (Embraer and Peruvian pilots). Among

EMB-312 FAP 431 taxies out for a patrol sortie close to the frontier. It is armed with two Heavy Machine Pods carrying 12.7mm (0.50in) machine guns. (Chris Lofting)

EMB-312 FAP 431 and FAP 432 in training colours during an instruction mission near Pisco Air Base. (Fuerza Aérea del Peru)

the weapons tested were the 70mm (2.75in) launchers with seven rocket each, SBAT 37mm (1.45in) with seven rockets for training purposes, FN Herstal Twin MAG Pod with two 7.62mm (.30in) MAG 58P machine guns, FN Herstal HMP Pod with one 12.7mm (.50in) machine gun; BEX-11 practice bombs (or Mk 76 Mod.5 of 11kg - 24lb) with smoke cartiridges and Macchi M910 pods with one 12.7mm (.50in) Browning M3 machine gun and 350 rounds of ammunition.

Flights were made at various FAPe regular operating sites, including bases ranging from 2,743m (9,000ft) to 4,572m (15,000ft) in altitude. In Laguna Salinas a simulation of an operational deployment was performed by taking off at 4,114m (13,500ft) altitude. At the firing range at Vitor Air Base, at 1,524m (5,000ft), rocket launches were made. The most important of the demonstrations took place at the Ayacucho Air Base at 2,743m (9,000ft) with the launch of BINC-200 200kg (409lb) incendiary bombs, rocket and machine gun fire at simulated targets on a 100-meter high slope at 90° with the horizontal plane.

Tactical navigations were made at low altitudes, in valleys and canyons 100m deep and 100m wide (328ft).

All the results were very satisfactory and convinced the Peruvians that the Tucano would be an excellent platform for the missions that the country pleaded to execute.

Thus, 10 more planes were bought and received in 1991 for USD 20 million in a Vietnam-style camouflage scheme and a shark's mouth. Of the equipment tested were used the FN Herstal Twin MAG pod, rocket machine gun pod with a 12.7mm (.50in) machine gun and 250 rounds plus three 70mm (2.75in) unguided rockets, 70mm rocket launchers with seven tubes each, BINC-200 200kg (408lb) incendiary bombs with stabilizer fins, Mk 81 and Mk 82 bombs with (Snakeye) and without delay and BEX-11 practice bombs with smoke cartridges.

Almost in secrecy, Escuadrón Aéreo No. 514 (No 514 Squadron) was born, a fully operational and capable unit to fly combat missions from Pisco Air Base, serving mainly the VI FAPe Territorial Air Region in Pucallpa, which covers the Alto Huallaga area in the centre of the country, dominated by the Amazon rainforest. But in order to ensure that the drug dealers' flights would not go unpunished and taking advantage of Tucano's operational flexibility, the planes were moved to other FAP bases such as Santa Lucia Air Base in the Huallaga river valley, an area responsible for 60 per cent of the cocaine produced in Peru, which had dozens of clandestine airstrips. They also flew from Iquitos, Pucallpa, Tarapoto, Andoas, Trompeteros, Ciro Alegría, El Valor and Tingo Maria. Drug dealers began to adopt other strategies, such as several planes taking off simultaneously in different airstrips and taking opposite directions in order to saturate the FAPe reaction. They also flew very close to the treetops to avoid detection by ground radar.

But since the US had a direct interest in restraining these flights, in 1995 a Lockheed P-3 AEW&C was sent to Peru through the Air Bridge Denial Program, to expand searches for these illegal flights. After locating them, the P-3 activated and carried out all the vectorization for the EMB-312 and Cessna A-37B towards the airplanes. This program aimed to help countries such as Peru, Colombia and Ecuador, for example, to increase the efficiency of their flights against these types of aircraft.

This new doctrine hindered and negatively impacted the finances of the drug dealers, who adopted a new strategy to keep the illicit activity in full operation. The pilots flew from clandestine runways on the border of Brazil, thus keeping the shortest possible time in Peruvian airspace and preventing them from being intercepted by FAPe. In order to get to these airstrips, however, the drugs followed a long and time-consuming land or river route and the take-offs were made at night to make it even harder to see and intercept these planes.

Once again, the US entered the scene and modified FAPe EMB-312s for night vision goggles (NVG) operations. The changes were made mainly in the cockpit and also

Ground crew were trained to quickly rearm and refuel FAP Tucanos to return them to flight condition for operational sorties. (Chris Lofting)

in the camouflage. The previous pattern with different colours disrupted the view through the NVG. The solution was to adopt a more uniform dark grey colour that facilitated the Tucano to be seen by another aircraft whose crew was also using NVG.

It is curious to note that the Peruvian Tucano pilots only flew with codenames, did neither use identifications nor rank. They were allowed to have beards and long hair. All aircraft at the flight line were always armed, even those that were not on alert.

From 1995 to 2001, FAPe operated extensively in the country's airspace looking to intercept these aircraft and the results were impressive. Under the co-ordination of the US AEW&C aircraft alone 14 planes were shot down by the Tucano and A-37, but the country says that the number reaches 40 with half being by the EMB-312 alone. Coca cultivation fell from 115,300 hectares (445 sq miles) in 1995 to 34,200 hectares (132 sq miles) in 2000.

But a serious mistake led to the immediate suspension of activity by the country. On 20 April 2001, a Cessna 185 seaplane with Peruvian civil registration OB-1408 carrying two Christian missionaries from the Association of Baptists for World Evangelism, their two sons and the pilot, was fired upon by a FAPe A-37C at 10.45. A woman, Veronica Bowers and her seven-month-old daughter, Charity, were killed. Her husband, Jim Bowers and her son were not injured and the pilot, Kevin Donaldson, was wounded in one of his legs by a bullet, but managed to land on the Amazon River, near Pebas.

A series of mistakes led the crews of the Cessna Citation SR-560 (which did the vectoring of the combat aircraft) and the two A-37B to that fateful tragedy. First of all, the Cessna 185 did not respond to radio calls since it was on HF because its VHF equipment presented problems. Another problem is that drug dealers often used seaplanes for this type of flight.

In spite of this the aircraft did not show any suspicious attitude like flying close to the trees. On the contrary, it was at 1,220m (4,000ft). The A-37Bs did not show up for the pilot so he would know he had been intercepted and given a chance to respond to the communication. The A-37 soon began the shoot down process. Thus, the FAPe was accused of negligence for failing to ascertain whether the Cessna carrying the missionaries was even hostile. As a consequence, there was a suspension on this type of flight in 2001 and the program was later discontinued in Peru.

The war in Cenepa

Ecuador and Peru have an unstable diplomatic relationship that has lasted since the 19th century. But it was only in the following century that the friction between these nations widened around the Cordillera del Condor. This is because, at the time of the independence of the Spanish colonies of that region – which included Peru, Ecuador and Colombia – the new nations agreed to define their territorial limits based on the principle of 'uti possidetis juris', that is, the borders would be the same as when they were colonies.

But Peru never completely accepted this definition and questioned some points in the Cordillera del Condor region.

In mid-December 1994, the Cordillera del Condor came to focus again and countries began to escalate with troop readiness, intensified training, patrol flights and border mobilization.

On 9 January 1995, in a direct confrontation, a patrol of four Peruvian soldiers from the 25° Batalhao Callao (25th Callao Battalion) was surprised and captured by men from the 63° Batalhao Gualaquiza (63rd Gualaquiza Battalion), Ecuador. The Peruvians, captured on their own territory, were disarmed and released hours later, but the incident caused Peru to send another patrol composed of 10 soldiers to the same spot two days later. Thus, at about 13.00 on 11 January, the Ecuadorians again surprised the Peruvians, who reacted by opening heavy fire, but eventually left the scene, leaving their armaments behind.

The FAPe, of course, had all its operational strength and attention focused on the war. And the EMB-312 were involved in this conflict.

On 4 February 1995, eight T-27s were on standby and armed with bombs to carry out an attack mission against Ecuadorian artillery, which in turn was commanding attacks on Peruvian troops near the town of Condor Mirador. The takeoff from Andoas aerodrome was around 01.00, using NVGs. '*On Saturday they knocked on the door of my dorm calling me to perform a night air raid mission. I put on my flight suit, and went mentalizing the route to the river Cenepa imagining the type of target to be attacked and its exact location. The photographs made by our personnel did not show where the target was and this artillery had already knocked down two helicopters, one from the Army and another from the FAPe*', recalls FAPe Lieutenant Carlos Figueroa.

At 01.30 all were in the briefing room ready to receive the mission instructions and the latest information obtained. '*The squadron commander entered the room. His face was tense, but his gaze was confident. He advised that after the attack we should return to Iquitos avoiding a possible counter-attack from the Ecuadorian Air Force with its IAI Kfir fighters. This entailed making navigation over the jungle*', he explains. It would be a journey of more than an hour in return.

There were eight pilots in total, two per aircraft. All received the latest weather forecast which indicated overcast sky, thunder and rain in some route points to the target. For the use of NVGs this was the worst case scenario for the absence of the stars or the moon creates a view with noise in the NVG lenses making the flight dangerous.

'*Leaving the briefing room we found the mechanics who wished us luck. I felt for the first time in life, a mixture of emotions, including anxiety and satisfaction. I climbed on the plane and my small flashlight lit the cockpit and seat belts so that I could prepare. Everything was ready. The radio remained off because we would not use it in flight. I cranked up the engine and its parameters were normal. With the flashlight I signaled that I was ready to take off. There were no lights. The track was cleared. I turned on the NVG, did the taxi and took off at full power*', he recalls.

The flight was made at 200m (656ft) high over the treetops. There was a lot of fog that morning, typical of the jungle region, reducing the horizontal visibility. As the targets approached the lightning storm worsened. '*We used each lightning to try to see some of the ground. The turbulence indicated that we were very close to great cloud formations, those that we should avoid at all costs. The flight became fully instrument. With five minutes to the target we went up a little more to avoid the mountain. The fog was down and we could see the area and where we should join and dive down to attack. I wondered if we had made a precise navigation. Any mistake meant a front and fatal collision against the mountains. I was listening to the wheezing of my co-pilot. I realized that I was even worse. It was when I put the oxygen of the aircraft at 100 per cent.*'

The storm was next to the aircraft and thunder ended up covering up the sound of the engines. 15 seconds from the attack the pilots began to see the target, albeit tenuously. Other visual references were used to confirm our location.

'I *told my co-pilot: now! Then I manoeuvreed and saw I was on the target. Synchronized the sights with the speed and height of the point of attack, then dropped four Mk 82 bombs. All went without any problem, as the weapons panel indicated by lights.*'

The return was made following a route parallel to the border. There was no counter attack or interception by the Kfirs. The pilots flew to their home base, to landing in Iquitos, after a 1h30min return trip.

The 2000s

The beginning of the new century severely impacted the EMB-312's life which almost had its operation suspended. The reasons were many, particularly the lack of maintenance and spare parts.

To work around the situation FAPe chose to put six cells on sale that were stored and which would stricken off the air force. All had around 50 per cent of useful operating life ahead and the values obtained with this sale, around USD 4.8 million, would be reversed for the purchase of spare parts to put the rest of the fleet in operation again. Angola proved to be interested in the deal, but the transaction was put on hold by the 13th Criminal Court of Lima, which reported that in early 2002 30 FAPe officers were involved in the irregular sale of assets belonging to the nation. In April 2003, Judge M. G. Morote personally inspected and filed the retention of six EMB-312, enacting judicial immobilization of cells sold to the Angolan government. Investigations took two years to complete and the cells were released in 2004 to the government of Angola.

Of the 30 Tucano originally purchased by Peru, five were lost in accidents and six sold, leaving 19 cells in the country. Of the damaged cells, one was recovered and transformed into a flight simulator. With the purchase of the KAI KT-1 advanced training and light attack turboprop, the Tucano gradually left advanced instruction in FAPe and in 2017 was in the final stage of decommissioning.

Left: During evaluation trials as an armed platform, this Peruvian EMB-312 was armed with an FN Herstal 12.7mm (0.5in) machine gun pod and a BEX-11 (11kg/24lb) exercise bomb. (Mauro César Mezzacappa)

Right: Another configuration tested comprised an FN Herstal Twin MAG Pod and BEX-11 exercise bomb. (Mauro César Mezzacappa)

UNITED KINGDOM

Royal Air Force (RAF)

The RAF is the most traditional air force on the planet. Maybe that is why the Embraer contract with the United Kingdom has been so important because in addition to ensuring an excellent deal it allowed the Brazilian company to learn more about the doctrine, needs and requirements of an operator so eminent.

Despite all efforts, the introduction of the RAF Tucano was not calm. On the contrary. Delays in the development program of the RAF version to be built by Short Brothers in Belfast, guarding profound differences compared to the original EMB-312, forced RAF to prolong the life of its Jet Provost fleet for another few years resulting in higher maintenance and operating costs for the fuel consumption of the Jet Provost was far superior and getting spare parts was naturally more difficult.

Also, because they are very different models and with little commonality of parts, RAF could not absorb the FAB experience and other military operators flying the EMB-312. Thus, the availability rates at the beginning were lower than expected until the initial problems, natural when introducing a new airplane in any air force in the world, were healed and dominated by RAF technicians and mechanics.

We can not leave aside also the criticisms and suspicions raised about Short Tucano turboprop because it was a tandem. There was a common consensus, as well as in Brazil during the development phase and even at the beginning of the EMB-312 operation, that the side-by-side configuration facilitated instruction once the instructor sees more

The flight line at Linton-on-Ouse is headed up by ZF170 from the 1st Flying Training School. (Derek Bower)

Wearing the traditional RAF training scheme of red and white, ZF511 belonged to the Empire Test Pilots' School and was used for test and training purposes at Boscombe Down. (Joop de Groot)

easily the student errors. Likewise, the cadet sees more easily their mistakes and can still look at the movements of the instructor and correct their faults.

But the introduction into service allowed the Short Tucano to quickly show its advantages.

Maintenance was much simpler to be made, much cheaper, the equipment was new and modern and far from the natural obsolescence that the Jet Provost had reached. Fuel consumption was three times lower. And its own flight characteristics and manoeuvrability won the instructors and students.

As with any new aircraft, RAF established a working group which was responsible for program development and training steps that the Tucano would perform with the cadets. It was necessary to see what was possible to adapt from the Jet Provost, what would be modified, deleted, replaced and even manoeuvres which could be included with the Tucano. All this work, long, complex and very detailed, was made by the Tucano Course Design Team at the Central Flying School. The first aircraft used was the ZF138 airframe which arrived for RAF at Scampton Air Base on 16 June 1988. This development work training program lasted until November 1989 and in December, at 7th FTS (Flying Training School) in Church Fenton, No. 42 Basic Flying Course became the first group to fly with the Short Tucano T.Mk 1.

The paint scheme adopted initially followed the pattern of the RAF training colours, namely white and red.

Gradually the Tucano flooded the flight line of RAF training units. The 3rd FTS based on Cranwell received ZF114 on 15 November 1990. The 6th FTS received on April 6 at its headquarters in Finningley the first Short Tucano that it worked at Tucano Air Navigation Squadron, a unit that was training all RAF navigators, so that these people could follow their training in rotary wing, transport or fighter aviation.

In January 1993, the Empire Test Pilots' School (ETPS), unit that trained pilots and flight engineers, received the last two Tucanos produced for the RAF – ZF510 and

RAF Tucano T.Mk 1s ZF138 and ZF139 in formation in the UK. (Archive Mario Vinagre)

ZF511, which just before had gone through the A&AEE (Aeroplane and Armament Experimental Establishment) at Boscombe Down. At ETPS trials and lateral, directional, stall and spin tests were performed. Both planes were equipped with telemetry, Automatic Voice Alert Device, an Air Data Acquisition Module, position indicators and assisted controls in both pilot stations and a probe at the tip of the left wing with skid and angle of attack indication fins.

In the first five years of operation RAF Tucano fleet had flown no less than 72,000 hours ensuring an average of 98 per cent availability performing up to four missions per day with one flight hour each. It is estimated that the fleet would have already exceeded 525,000 hours of flight at RAF until the end of 2017.

In 1993, the most flown RAF aircraft had almost reached its structural lifespan, reaching 11,000 of 12,000 planned hours. To maintain this availability, the RAF kept in the mid-1990s around 75 aircraft in the squadrons, while otherswere stored or at maintenance in Shawbury. This meant that the availability and control of the life of the aircraft was made more rigorously, allowing in general that the fleet could maintain a use and consumption of flight hours similar and equal.

The end of the Cold War and the impact on the training of RAF

As in France, the end of the Cold War impacted in a drastic reduction of military expenditures and also the demand for large quantities of pilots since the air units were smaller and fewer planes and flight hours.

Thus RAF was forced to undertake a profound change in the system for training and formation of its pilots.

In 1992, 7th FTS ceased to exist and its Tucano fleet was fully transferred to the 1st FTS in Linton-on-Ouse, which already at that time was receiving the first units of the Tucano divided into two squadrons. Gradually, the 3rd FTS also transferred its planes to 1st FTS. In 1995, the 1st FTS also incorporated the Central Flying School, which in turn operated for a brief period off Topcliffe Base, near Linton-or-Dare. The Tucano Air Navigation Squadron of the 6th FTS in Finningley, went for the 1st FTS concentrating its activities in Linton-on-Ouse.

In 1997, the Central Flying School went to Linton-on-Ouse, followed by the Tucano Air Navigation Squadron in 2003. In 2007, Tucano Air Navigation Squadron turned into 76 (Reserve) Squadron until 2011, when it was disbanded.

Throughout its operation at RAF, Tucano composed the transition to the fighter, interdiction and attack jet aircraft in the basic fast jet training program. Thus, completing basic training the students who would follow to fighter and attack aviation passed by Tucano while making basic fast jet training. It lasted 10 months (40 weeks) with 150 flight hours. The first weeks are dedicated to converting the students to the Tucano (ground school), where flight simulators are used used, learning basic manouvers and Tucano's general handling, as well as normal and emergency procedures.

The first stage of flight instruction aims at basic training manoeuvres, basic and instrument navigation, aerobatic manoeuvres, low altitude navigation, high-perfor-

The Short Tucano was a key asset in the training course for future RAF pilots. (Derek Bower)

ZF142 from the 1st Flying Training School at RAF Linton-on-Ouse displays the gloss black scheme adopted to increase conspicuousness at greater distances. (Neil Bury)

mance manoeuvres and emergency procedures. This first stage consumes about half of the flights scheduled for the course. The first solo flight was done after the 10th flight, with 30 minutes long on basic closed circuit. During the course, the student will face around 26 hours of solo flights. Aerobatic flights were also done, including tounneaux and loopings using a base altitude of 1,524m (5,000 feet).

In the second half of the course the emphasis is given to the use of tactical aircraft training pilot for future combat missions. In this step are carried out flight missions in formation, night flying, advanced navigation, tactical navigation, attack simulation, tactical flights at low altitude and long-distance navigation from 914m to 1,219m (3,000 and 4,000 ft) altitude. Bad wheather flights were also done as navigation without the use of GPS, only compass, map and chronometer, looking always outside of the cockpit. Attack simulation include to track, locate and attack the target at the desired and planned time. After 60 hours, it start the low level navigation and there were also formation flights with 3 meters (10 feet) separation between each plane. Its include levelled flight, curves and then breaks and the introduction of IFR day-time flights. The next step is five night IFR flights.

After a change in the aircraft, which began to incorporate the Traffic Collision Avoidance System (TCAS), the Tucano began to conduct training for use of radar using this system since the TCAS informs the relative altitude and azimuth of the other Tucano aircraft so equipped. After the Basic Jet Fast Training, pilots follow to the advanced tactical course at the 208 (Res) Squadron flying the Hawk T.1.

RAF, after conducting a series of studies concluded that the white and red paint scheme of its planes was more difficult to see in flight and that this pattern could even cause accidents. Thus, it was decided to change the colour scheme for the glossy black with numbers, markings and warnings in white. While that was much more visible when it was farther away from other aircraft in flight, it has become one of the most beautiful and aggressive training paintings that Tucano ever used worldwide.

In mid-2000 the fleet was completely grounded after being identified above normal wear on the rudder commands so that all the planes went through overhauls and replacement of these components with new and more strengthened ones.

In addition, RAF was also forced to add structural reinforcements in the structure of aircraft including wing spars and the attachment points of these with the fuselage.

The arrival of the 2000s also created other less significant changes in training with the Tucano. The two squadrons of the 1 FTS were renamed 72 and 207 (Reserve) Squadron, respectively.

In 2007, the number of Tucano in operation fell to 55, with RAF going to use the part of the fleet that was more worn or had reached greater amount of flying hours as a source for spare parts for the ones still on the flightline. In addition, another 30 aircraft were put for sale by the UK Ministry of Defence through the Disposal Services Agency. Many were no longer in flight, but part of these could return to flying condition if investments were made by their future owners. One of them, ZF373, was registered as G-CEHJ, came to be serviced in Harwarden, but was not put into flight again. Another six aircraft were exported to Oregon, US, receiving US civilian registration numbers N8093S, N8093Z, N1425S, N192S, N1948S and N710RS.

Until July 2017, 82 airframes were demobbed. Fourty three were moved to US. Twenty-nine more are still stored in UK, one in a museum and an other one preserved. There are anotter six used for spare parts with two in private use in UK.

The future

Over the past 16 years, the UK has conducted studies to reduce training costs and optimize this mission in the country's three armed forces. One of the conclusions pointed to the unification of training systems and privatization so the companies would provide training management, maintenance, competition to buy flight simulators and dedicated aircraft and provision of minimum quantities of airframes and flight hours. The Ministry of Defence would only provide instructors, fuel and air bases.

Called Military Flying Training System, the Ascent consortium of Lockheed Martin UK and VT Group won the management of the unified training program for a period of 25 years with an estimated investment of USD 7.6 billion. The contract was signed in 2008 and in that year the operation of the Tucano, scheduled to end in 2010, was expanded to 2012. Other maintenance and support contracts delayed the withdrawal of Tucano fleet for 2014 and later to 2017.

In December 2016 10 Beechcraft T-6C Texan II were selected through the Affinity Flying Services consortium, a joint venture between Elbit Systems/Killogg Brown and Root Ltd (KBR), to finally replace the Tucano in RAF service. In addition to the T-6C Texan II, 23 Grob G120T and five Embraer Phenom 100 were choosen, covering from elementary instruction with the Grob, basic with the Texan II and multi-engined with the Phenom 100. It is expected that the Short Tucano will be withdrawn from operation in 2019 or 2020.

Tucano T.Mk 1 ZF287 from No. 72 (Reserve) Squadron in action in one of the UK's low-flying areas. (Michael Brazier)

UNITED STATES

Private owners

France's EMB-312F and the RAF Short S-312 T. Mk 1 Tucano have gained a second operational life in the hands of US private civilian operators. French airframes were the ones with the greater remaining lifespan. Because they were prematurely withdrawn from service, France consumed on average only one-third of the available flight hours before their cells reached the operational fatigue limit.

As for the RAF airframes, they had higher consumption of flight hours as the aircraft entered active service in the United Kingdom in the late 1980s and remains flying to this day.

However, by 2017, 24 Short Tucano and 15 EMB-312F had been registered in the United States. Several reasons lead to understand what those planes were doing in the USA.

The first batch arrived on American soil in 2008. Those were six Short Tucano acquired from the RAF partially disassembled and all with original equipment.

The customer was an operator in Oregon, who for a while kept the aircraft dismantled. Naturally, with the passage of time, some interested people emerged, the aircraft were assembled and gradually sold to be used for leisure and sports.

On 22 June 2015, one of those aircraft, the ex-RAF ZF206, crashed and killing the renowned composer James Horner, who composed the soundtracks of Titanic and Avatar, two Hollywood blockbusters.

As these airframes came into operation they soon made their characteristics and flexibility spread throughout the United States.

Thus, civilian companies that own military aircraft to promote specialized training for the US Armed Forces began to look at the Tucano differently. Despite being a turboprop it was fast and manoeuvrable at different flight levels. It was easy to maintain and operate, versatile and its cost of acquisition was low (around USD 350,000). And

This privately owned EMB-312F, formerly Armée de l'Air serial number 484, was exported to Military Tech Inc. in the US. The role of the underwing pods is unknown and the aircraft was last seen in 2016. (Jack Mcgorray)

Many private operators have adopted Short Tucano T.Mk 1s, after their retirement by the RAF. ZF170 is seen at the Oshkosh Airshow. (Andreas Zeitler)

the architecture of the cockpit allowed the turboprop to conduct training as if the pilot was flying in a jet.

It was the perfect world and so other operators started to import other RAF Tucano that were for sale.

RSW Aviation based in Phoenix, Arizona, specializes in providing advanced training for pilots to respond to extreme, adverse situations that require a quick and accurate response avoiding an accident. The classic example is turbulence, flaws in aircraft systems, flight in adverse weather conditions and other situations.

Over time, the company began to launch a modernization program for the Short Tucano by installing, among other things, a digital panel with two coloured multifunction liquid crystal displays certified for IFR and other more modern components to reduce the obsolescence of the aircraft.

Some other operators opted for the Garmin 600 avionics with a Primary Flight Display, a Multi-Function Display and a Garmin GTN750 touch screen panel.

In 2016, RSW Aviation further advanced the exploration of other activities with Tucano. This time, it certified a structural shift in the wings to allow the attachment of four pylons carrying a wide range of weapons such as launchers with conventional (non-guided) 70mm (2.75in) rockets and even with the laser-guided Advanced Precision Kill Weapon System. In addition, up to two racks to carry two AGM-114 Hellfire air-to-ground missiles each, up to two Heavy Machinegun Pod (HMP) with a 12.7mm (.50in) FN Herstal M3P machine gun or two FN Herstal Rocket Machine Gun Pod with a 12.7mm (.50in) M3P machine gun with 250 rounds plus three 70mm (2.75in) unguided rockets.

RSW Aviation included in the aircraft consoles that allowed the installation of monocular sights mounted on the helmet. In addition, it made a change in the exhaust so that the heat flow was directed to the sides and under the wings instead of following the flow in the ventral direction. Thus, it was possible to install an L3/Wescam MX-15 infrared camera just below the cockpit without the heat of the exhaust influencing the system.

Currently the company has 11 aircraft, almost half of the US registered fleet.

Military Tech INC operates seven EMB-312F and two Short Tucano, and little is known about what the company's mission has been for these turboprops. Only in January 2017, the civil registered EMB-312F N484MT was spotted in flight. The aircraft had a two-tone grey camouflage pattern and only the black inscription of its registry. However, under each of the wings, an operational training designator can be seen. The other aircraft of the fleet, both EMB-312F and S-312 T Mk 1, remain in secrecy.

In addition to sporting and private military training for the US Armed Forces and even for pilots from other countries, experts say the purpose of these companies is to sell some of these aircraft to small African and Central American Air Forces. But so far no sale has been confirmed.

Some US companies have offered modernisation services for the Short Tucano including glass cockpit panels, like this Garmin 600 in ZF200. (Advantage Avionics)

VENEZUELA

Fuerza Aérea Venezolana (FAV, 1946–2007)

Aviación Militar Bolivariana (AMB, 2007 until today)

While Honduras was Tucano's first export customer, the Venezuelan Air Force was the first nation in South America to choose Tucano as part of the training of its future military pilots.

In 1985, through a contract valued at USD 51 million, Venezuela acquired 30 aircraft, spare parts and specific training of pilots and mechanics. On 13 July 1986, the first four copies left Embraer in São José dos Campos to the Generalísimo Francisco de Miranda Air Base in Caracas, where they were received at 17.00 the following day after a long ferry flight. At the grand reception ceremony the President of Venezuela, Jaime Lusinchi, Defence Minister General José Rafael Cardozo, the General Aviation Commander of the Country, Major General Ramón Enrique and other civilian and military authorities attended.

The deliveries continued until 1988, always in batches of four airplanes.

The first 20 were delivered in an aluminum-coloured pattern with a red stripe applied to the wing tips and fuselage – this with the letter 'E' in white indicating that the aircraft was in use by the Escuela de Aviación Militar (Military Aviation School). The aircraft were assigned to the Grupo de Entrenamiento Aéreo No. 14 (GEA, Aviation Training Group), the 'Escorpiones', based in Mariscal Sucre – Maracay with the designation T-27.

Wearing brand-new camouflage, EMB-312 FAV 0175 is armed with two seven-tube 70mm (2.75in) rocket pods and two pylons for four BEX-11 (11kg/24lb) exercise bombs. (Ivan Peña Nesbit)

Holding a sign displaying the 134,0 MHz emergency radio frequency, FAV 1405 takes part in a joint interception flight with the Brazilian Air Force. (Paulo Rezende, FAB)

In that group, the future military pilot did his primary training in the Beechraft T-34A Mentor, in the Escuadrón de Entrenamiento Primário No. 141 (No. 141 Primary Training Squadron). He then moved on to the Escuadrón de Entrenamiento Avanzado No. 142 (No. 142 Advanced Training Squadron) where the Tucano would fly IFR formation flights, advanced navigation, night flights and aerobatics.

In the Fuerza Aérea Venezolana (FAV, Venezulean Air Force) the EMB-312 replaced the obsolete and costly BAC Jet Provost.

Two AMB EMB-312s armed with machine guns and rocket pods. (Ivan P. Nesbit)

FAV 0086, FAV 0214 and FAV 0100 (front to back) in the silver training colour scheme overfly Venezuelan territory. (Fuerza Aérea Venezoelana)

Tactical support missions

With the arrival of the first AT-27 Tucano, in August 1987, the Escuadrón 152 'Avispones' was created, belonging to the Grupo Aéreo de Operaciones Especiales No. 15 'Zorros', located at Rafael Urdaneta Air Base in Maracaibo. These planes bore a tactical painting of jungle camouflage in dark green, medium green and tan, and also using the red colour on the wing tips and the fuselage, but without the letter 'E'. At their disposal were LAU-32 70mm (2.75 in) rocket launchers with seven tubes each, Heavy Machine Pod (HMP) machine gun pods with a 12.7mm (.50) FN Herstal M3P machine gun or FN Herstal Rocket Machine Gun Pod with a 12.7mm (.50) M3P machine gun with 250 rounds plus three 70mm (2.75inches) unguided rockets. In addition it used Mk 81 and Mk 82 bombs.

This group had the Rockwell OV-10 Bronco multipurpose units assigned to the Escuadrón No. 151 and performed a wide range of missions such as escort, patrol, close air support, ground attack, interception of small aircraft at the service of drug dealers, armed reconnaissance and others. To date, no aircraft have been shot down by FAV EMB-312, but several were forced to land for investigation and others were taken out of Venezuelan airspace. Airplanes were often deployed for operations from Puerto Ayacucho Air Base in southern Venezuela, where they acted against guerrillas.

Reorganisations and changes

In 1991, the Escuadrón 152 gave up its EMB-312 to start operating the new OV-10 Bronco batches that the FAV had bought from the USA.

These aircraft were then relocated to the Escuadrón No. 131 'Leones' (131 Squadron), belonging to Grupo de Entrenamiento Táctico No. 13 (No. 13 Tactical Training

Group)and based at Luis Del Valle Garcia Air Base – Barcelona. This unit, formerly known as Grupo de Bombardero No. 13, flew the legendary BAC Canberra.

After flying the T-34A Mentor and Tucano, those students who would follow in the instruction of combat pilots, such as fighter and attack, flew the North American T-2D Buckeye receiving in the Escuadrón No. 131 'Leones' the doctrines of aerial interception and ground attack with even more advanced manoeuvres. But already in the early 1990s these planes suffered with the low availability and Tucano came precisely to reinforce this task in the FAV.

EMB-312 FAV 0100 after a forced landing. Minor damage can be seen on the aircraft's belly. It was later hoisted away by a Bell 214.
(Archive João Paulo Moralez)

In the early 2000s the T-34A were replaced by the Aermacchi SF-260E in Escuadrón No. 141. The T-2D Buckeye also left service on the Esquadrón No. 131 which was then deactivated and transferred their EMB-312 to GEA No. 14.

As well as Escuadrón No. 141 now with the SF-260E in primary instruction and Escuadrón No. 142 in advanced instruction with the EMB-312, was created the Esquadrón No. 143 of tactical training in the GEA No. 14 also with the EMB-312 Tucano that belonged to the Escuadrón No. 131. This new squadron started to perform the same missions as the T-2D Buckeye – that is, training the combat pilots for fighter and attack missions.

There were also changes in the Grupo Aéreo de Operaciones Especiales No. 15. Also in the early 2000s FAV suffered from the low availability of its OV-10 Bronco fleet. To increase the number of airplanes in flight conditions, all OV-10s became concentrated in the Escuadrón No. 151, with some EMB-312 being brought back to the Escuadrón No. 152 'Avispones', returning to the same organization of the 1980s.

On 12 December 1987, FAV lost its first T-27, however without the pilot suffering injuries after ejecting successfully. The airframe was replaced, and the number of EMB-312 in Venezuela elevated to 31.

It is curious to note that in FAV the military registration of aircraft did not follow a logical sequence. Upon arrival at FAV, an already numbered card (four digits), filled in with the airplane data, determines what your registration will be in the country. Thus the aircraft received the numbers 0040, 0086, 0100, 0150, 0175, 0203, 0214, 0260, 1335, 1405, 1792, 2170, 2370, 2460, 2820, 3050, 3105, 3131, 4000, 4180, 5050, 5613, 5625, 6300, 6888, 7101, 7321, 8090, 8247, 9153 and 9244. FAV would lose another five units in accidents.

On 25 May 2001, a mid-air collision between FAV 1792 and FAV 9153 resulted in total loss on both aircraft yet the four crew members involved, managed to eject successfully. On 9 December 19 of that year, FAV 3105 crashed into Cerro Carmelo followed by FAV 2170 on 20 January 2004. On 22 March 2005, it was the turn of FAV 8090.

Fighting in 1992

The year 1992 was extremely troubled for Venezuela due to two attempts of coups d'etat in the country. In both situations the actions were triggered by left-wing politicians, self-styled anti-imperialist Bolivarians, honoring the leader Simón José Antonio de la Santísima Trinidad Bolívar and Palacios who liberated Venezuela, Bolivia, Colombia, Ecuador, Peru and Venezuela in the 19th century from dependence and the Spanish mandate.

And FAV's EMB-312 Tucanos performed in both situations.

A close-up of FAV 5625's tail section, clearly showing the combat damage. (Archive João Paulo Moralez)

On the night of 4 February, the respected Lieutenant Colonel Hugo Rafael Cháves Frias, of the Venezuelan Army Parachute Regiment, led the coup attempt to the doors of the Presidential Palace. But almost every contingent of the armed forces remained loyal to the government of Carlos Andrés Perez, who fled through a tunnel, managing to stifle the movement.

Support for the rebels came mainly from the Army and Air Force, with insurgents capturing a few EMB-312s and some Lockheed F-16As. Only a few flights were made and no plane was actually used in combat since the attempted coup ended on 5 February with the arrest of its top leaders.

Hugo Chávez, himself in prison, planned the second coup, which was much more organized than the first one.

Taking advantage of the Air Force Day festivities traditionally celebrated on 10 December with static displays and air shows, the conspirators gathered at one of the most strategic and largest FAV bases, the El Libertador Air Base in Palo Negro, home to a large part of the air force's combat fleet. At dawn on 27 November 1992, at 03.30, several radio, television and military units were taken by the rebels, including Escuela de Aviácion Militar located at Mariscal Sucre Air Base where many EMB-312s were based and some T-2 Buckeyes were located. El Libertador Air Force Base also fell into the hands of the rebels. Meanwhile, the F-16As on alert were evacuated by government-loyal pilots who flew to Teniente Vicente Landaeta Gil Air Base in Barquisimeto joining the Northrop F-5s of Grupo Aéreo de Caza No. 12 and also to some T-2s that were on the site.

As soon as the day was born, at 06.15 two Dassault Mirage 50EVs attacked Army facilities at Fort Tiuna causing casualties in western Caracas. Moments later another four OV-10 Broncos, two T-2s and six Tucano AT-27s began attacks on the Presidential Palace and the Foreign Ministry Building in Caracas with Mk 81 bombs and 70mm rockets. All these units and barracks remained loyal to President Carlos Andrés Perez.

At 13.00, the Broncos and Tucanos returned to attack Caracas, mainly Generalísimo Francisco de Miranda Air Base. But the governist force was able to reorganize and counter-attack by advancing on the rebels in Caracas and on that occasion one of the governist F-16A under Captain Labarca's command attempted to attack the OV-10 Broncos and EMB-312s. However, over the capital Caracas, flying very low, with many buildings, mountains and a huge speed difference between planes, the F-16 was not able to shoot down any rebel aircraft.

Soon the F-16As dominated the airspace while in the ground Carlos Andrés Perez's Army managed to reverse the situation. The rebels, for their part, were unable to put any of the F-16s captured in El Libertador in flight, thus demonstrating the fragility of keeping the attack flights then dominated by the government.

The EMB-312 flew on Forward Air Control missions to ensure the safety perimeter of El Libertador and Mariscal Sucre. They were armed with the Heavy Machine Pod (HMP) and even attacked armored vehicles and columns of soldiers trying to approach the bases.

The movement was once again losing strength. At 15.00, the government proceeded to attack the rebels' stronghold at El Libertador with the F-16s, while on the ground the base was almost encircled. At that moment Lockheed C-130F FAV 2715 took off taking the rebels to Peru. The second coup attempt was over.

Both the OV-10 Bronco and the EMB-312 were shot at by various anti-aircraft artillery gauges, such as 7.62mm (.30), 12.7mm (.50), 20mm and 40mm.

Virtually all T-27s and AT-27s had wing and fuselage damage from artillery shots. But none were shot down and two crash landed after suffering engine damage. One of them had two pilots. One was killed by government troops and another died at the hospital as a result of gunshot wounds sustained in flight and inside the aircraft cockpit.

It is worth mentioning that the rebels also used the Tucano to flee to neighbouring countries as the government regained control of the country.

Current service

From 2014 onwards, FAV changed once again its structure. With the acquisition of the Diamond DA40, this aircraft now provides primary and basic instruction for future FAV military pilots. Those who will go on to transport aviation are sent to fly the Diamond DA42. The pilots selected for combat aviation continue to GEA No. 14 to fly the SF-260E, Escuadrón de Entrenamiento Primário No. 141, and the EMB-312 in the basic and intermediate instruction of the combat pilot, at the Escuadrón de Entrenamiento Avanzado No. 142 and Escuadrón de Entrenamiento Táctico No. 143.

A flight line of armed EMB-312s from Grupo de Entrenamiento Aéreo No. 14 during exercise sorties. (Ivan P. Nesbit)

AFTER TUCANO AND THE ROADMAP TO SUPER TUCANO

Super Tucano

The development of Tucano for RAF in partnership with Short Brothers made Embraer review the concepts of its original trainer, which led to the realization that its product, without receiving any upgrades or improvements, was losing market shares to the latest variants of Pilatus, such as PC-9 and later PC-9M.

Pilatus' planes had incorporated ejection seats, the single lever, and a few other Tucano concepts with the advantage of having a much more powerful 950hp engine versus the Tucano's 750hp.

In order to reverse this situation, Embraer equipped PT-ZTW, one of its demonstrators, with a 1,100hp Garrett TPE 331-12B turboprop, four-bladed propeller and a diving brake making it similar to the Short version, the EMB-312G2. That way was born the EMB-312G1. PT-ZTW flew for the first time on 28 July 1986, but this variant did not go forward. The G1 would compete directly with Short Tucano, so the sales volume could be low not justifying the investment.

It is important to note that Embraer was working on the EMB-120 Brasilia regional turboprop; in a bi-national program with Argentina on a regional turboprop, the CBA-123; and the AMX attack jet, a bi-national program with Italy. It also came to the late 1980s facing its worst financial crisis, which had led to the firing of employees and overloading those who had stayed. So time, money, and human resources were scarce in the company.

But in February 1987, Embraer continue to study a second version, the EMB-312G1A, which would have a more reinforced structure, HUD and digital cockpit being intended for even more advanced training including air-to-air combat, low-altitude navigation, ground-attack missions and greater resistance to manoeuvres supporting superior G loads. This version would be offered to FAB, who was already thinking of using Tucano to equip the future combat units that would be created in Porto Velho and Boa Vista in the mid-1990s, and the 2ª ELO itself.

Another project that was studied, but not left out, was the helicopter-killer variant, which would be monoplace, lighter, with greater fuel capacity and weaponry (including missiles) and much more manoeuvrable at low heights and speed. Among the items to be included were a HUD, Kevlar cockpit armour (on the floor and sides); self-sealing tanks; laser rangefinder; fire-and-forget missiles, two 20mm cannons with a cadence of 4,000 rounds per minute and a 1,100-horsepower engine. Some tests were done with an EMB-312 and the modified 1,100hp Garret engined PT-ZTW against a Helibras HB-350 Esquilo from CTA, training and analyzing these dissimilar combats. But the program did not go forward.

EMB-314 Super Tucano prototype PT-ZTW. (Archive Col Aparecido Camazano Alamino)

But even if it had not decided to continue with these developments, versions G1 and helicopter-killer, the fact that France and the United Kingdom introduced the turboprop in the training and instruction of their military pilots led other nations to take a closer look at this growing niche on the international stage. These countries generally sought out more agile platforms with modern avionics and firepower for weapons training missions, or even the possibility of being used on anti-guerrilla missions or fighting in limited, localized conflicts. In some countries it could be used for interception of light aircraft.

By the end of the 1980s, G1A continued to be studied and evoluted to the EMB-312F. The model would be equipped with a 1,100hp Pratt & Whitney PT6A-67/13, four-bladed propeller and with Full Authority Digital Engine Control (FADEC), a fully digital control made by computers. Structurally it would support +8/-4g; cell with 15,000 planned flight hours; new leading edges and an even stronger canopy; pressurized cockpit (2.5 psi differential), digital, with HUD, new ejection seats and anti-g system for the pilots; aerodynamic brake and new avionics.

These were some of the characteristics described in the report PT-312F/879, mentioning the EMB-312F variant which was then treated as EMB-312H since the letter 'F' was destined for the French Tucano. Dimensions have not changed.

The PT-312H/962 report of September 1988 changed the engine to a 1,250hp PT6A-67/17 but retaining the propeller.

American dream

In the early 1990s the USAF and the US Navy were seeking to acquire a single trainer to jointly meet their demands. Of course, each one has its peculiarities in terms of doctrine, but in basic and intermediate training using a single-engine turboprop could meet the requirements of both for a much lower cost. In addition, the volume of purchases added to both forces would further reduce the operation over the decades. So, after each one established its requirements, the Joint Primary Aircraft Training System (JPATS) was launched.

After RAF, the US would be the glory for Embraer. In addition to being able to conquer new markets, a supply of airplanes to the US was always in the hundreds and with important contracts of sale of parts and maintenance.

With PT-ZTW, Embraer created a prototype and concept demonstrator proving that it could meet the requirements of JPATS if elected the winner. Internally the program was launched at Embraer in January 1991 and worldwide in June of that year in France at Le Bourget.

The aircraft would undergo a further series of modifications to receive a new 1,600hp Pratt & Whitney PT6A-67R engine and a five-blade Hartzell propeller; lengthening of the fuselage by 1.37m, the installation of a new segment in the front to house the engine and another in the rear to compensate for the centre of gravity (CG); installation of zero-zero type Martin-BakerMk 10 ejection seats; new supply system under pressure and On-board Oxygen Generating System (OBOGS). It had maximum take-off weight of 2,770kg (6,106lb) or 3,410kg (7,517lb) with external loads.

The changes were made very quickly and on 9 September 1991 PT-ZTW with these modifications and under the command of test pilot Gilberto Pedrosa Schittini and flight test engineer Mauro César Mezzacappa successfully held its inaugural flight. If the model was accepted Embraer would offer the JPATS aircraft with glass cockpit, reinforced structure and other technological advances of the time.

The EMB-312H received a beautiful blue paint with a yellow lightning bolt, which ran almost the entire length of its fuselage. At the time, a Brazilian television channel was showing the soap opera 'Ana Raio and Zé Trovão' – something like 'Lightning Ann and Joe Thunder' – which soon led Embraer employees to call it Ana Raio's plane.

Towards competition

As was the case in the United Kingdom, Embraer had to join an American company in order to participate in the competition and to produce the aircraft and provide all technical assistance, if it was the winner.

In May 1992 a protocol of intent was signed between Embraer, through its superintendent director, Ozires Silva, and Northrop vice president and general manager, Wallace Solberg.

Having baptized the EMB-312H as Super Tucano, Embraer sent PT-ZTW along with PP-ZTT – a conventional unmodified EMB-312 demonstrator – to the US on 27 July 1992. The flight departed from São José dos Campos with stops in Brasilia, Belém, Cayenne, Barbados, St. Thomas and Grand Turk. They landed at Embraer Aircraft Corporation, the company's subsidiary in Fort Lauderdale on the 30th.

Since then, until 7 August, they made training and promotional flights.

The next day they went to Andrews Air Force Base where they performed 38 demonstration flights until 22 August. On 23 August, they then proceeded to the Witting Field Air Force Base in Pensacola, Florida.

Interestingly to note that on that day, while the USN's Beechcraft T-34 were evacuated to other locations, as the city could be hit by Hurricane Andrew, the two EMB-312Hs made their demonstration as planned, leaving the city two days later and shortening his stay that should be seven days. From 27 August to 4 September, the planes performed at Corpus Christi Air Force Base and then stayed until 11 September at Randolph Air Force Base.

Altogether there were almost 100 hours of demonstrations on 83 flights.

In addition to Embraer, Pilatus and Beechcraft participated in the competition, offering PC-9 Mk 2; Cessna with the 526 CitationJet; Aermacchi with a MB-339 variant called T-Bird II, in association with Lockheed; SIAI Marchetti with S.211 associated with Grumman; Rockwell with Ranger 2000, based on the RFB Fantrainer and Vought with the Pampa 2000, a modified variant of the Argentine FMA IA-63 Pampa. Of all participants, only Embraer and Pilatus offered a turboprop trainer, while all the other were jets.

The end of a dream and a new opportunity

For Embraer, however, JPATS did not work. In June 1995, Pilatus partnership with Beechcraft won the competition to provide 372 units of the T-6 Texan II to the USAF and 339 to the USN, for a total of USD 4 billion. The cost of the fleet's life cycle can reach another USD 7 billion.

The loss of this billionaire contract further affected the financial crisis that was at its peak in 1995. It was a blow to Embraer, but it was able to take advantage of the dividends of this competition to offer to FAB a new product, much more advanced than Tucano itself. It would be an aircraft to fulfill Brazil's longing to patrol the Amazon region and the entire dry frontier with a robust, advanced and easy-to-operate platform with low acquisition and operating costs.

The Super Tucano had flight characteristics, embedded systems and engineering and technology solutions that made it virtually unique in the segment. With a digital cockpit, HUD, HOTAS (hands on throttle and stick) system, armour, greater weapons capability, capable of flying with night vision goggles, use smart weapons and simulate advanced stages of military pilot training.

That was what was missing for this segment and once again Embraer was one step ahead. The USAF is considering the Super Tucano as an option to assume the majority of its A-10 Thunderbolt II missions.

As of November 2017, the Super Tucano has been selected by 14 air forces worldwide. Other countries are also expected to announce orders in due course.

A forthcoming book from Harpia Publishing will fully describe this state-of-the-art combat aircraft, It will include details of how Embraer developed the Super Tucano, its characteristics, expanded weapons capability and why this 'Amazon warrior' is still conquering the world.

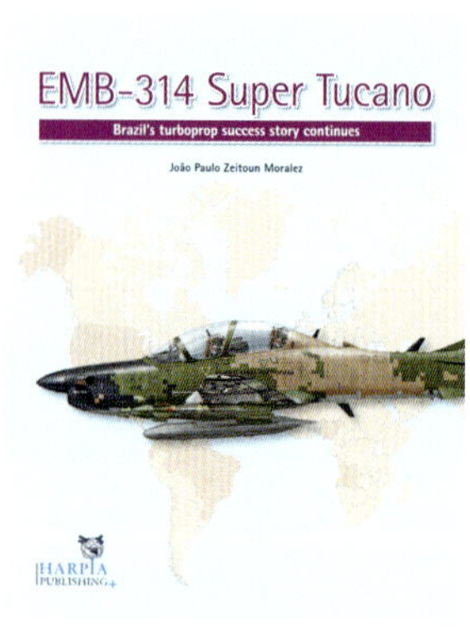

EMB-314 Super Tucano

APPENDIX I: TECHNICAL DATA

EMBRAER EMB-312 TUCANO

Manufacturer
Designed by Embraer in São José dos Campos (SP), Brazil. Project engineer was Joseph Kovács and the technical director was Guido Pessotti.

The inaugural flight of the first prototype was made on 16 August 1980 and the first six production aircraft delivered on 29 September 1983 to the Brazilian Air Force.

All variants are biplace. The aircraft were built in Hangar F-60, F-60/2 (attached to hangar F-60) and then F-107 (from 1985), all of them at São José dos Campos.

Egypt acquired the Tucano with the production right under licence from Kader Factory in Heliopolis. After being assembled they were transported by land to Helwan where they were painted and carried out the tests in flight and the official delivery. The contract with Egypt established that ten copies would be produced by Embraer; 42 kits of large sets such as wings, fuselages and empennage were made in Brazil and sent to Egypt; While 68 kits were boxed in Brazil and sent to be fully assembled in that Arab country. Subsequently another 14 additional kits were purchased. Of the total of 134 copies, 80 went to Iraq and 54 to Egypt. The Arab country produced 124 EMB-312 Tucano.

Production
Total of 475 airplanes were produced. Of this, 463 series, one prototype and one pre-production were delivered to Angola (14, six new ones, six acquired from Peru and two Embraer demonstrators); Argentina (30); Brazil (135 being 133 serial, one prototype FAB 1301 and one pre-production 1302); Colombia (14), Egypt (54), France (50), Honduras (12), Iran (15), Iraq (80), Mauritania (06, donated by France), Paraguay (09, with three bought from Brazil), Peru (30) and Venezuela (31).

One flying prototype, FAB 1300, was not delivered to Brazilian Air Force being scraped after. Three non-flying structural test cells did use c/n 312. Embraer also produced three demonstrators. The PP-ZTC, which was transformed into Short Tucano prototype G-BUTC; PP-ZTT and PT-ZTK (counted above), which was sold to Angola in 2003, mentioned above, receiving the military registrations R-456 and R-457.

There was also a crash loss of an EMB-312 during test flights before being delivered to FAB, the registration 1317. Another one was produced but refused by France, and then dismantled. There was also PT-ZTW, which was the EMB-312G1 prototype and then the EMB-312 H prototype. And the PT-ZTV and PT-ZTF, prototypes for JPATS, which used the c/n 312. There is also 27 more 312 c/n allocated to 27 kits produced by Embraer as Short Tucano for UK but still using c/n 312, totally 502 units.

Role

Basic training, advanced training, fighter pilot training with weaponry, ground attack and fighter. It is aerobatic and IFR capable.

Crew

In tandem and equipped with two ejection seats Martin-BakerMk 8L 0-70 (zero altitude and 70 knots – 130km/h). The instructor sits on a raised seat in the rear cockpit to observe the actions of the student sitting in the front seat. The take-off, when soloed, can only be made with the pilot occupying the front position of the aircraft.

Analog panel with dual control, monomannel for engine control, rear view mirrors and a conventional reflex sight DF Vasconcelos.

Airframe and systems

Canopy in single piece and without windshield. Air-conditioned, non-pressurized, 2024 aluminum alloy frame and cantilever wings and feather construction. Retractable tricycle landing gear with single tires and fully protected by doors made of composite material when collected.

All components of fuselage flight commands; The complete structure of the wings; The flaps; ailerons; stabilizer; elevators; rudder; landing gear; motor; engine crib;propeller; propeller spiner; exhausts; ejection seats; antennas; hydraulic accumulator; hydraulic reservoir and battery are interchangeable and without the need of modifications.

The wings have two fuel tanks each for 694 liters (183gal) in total. Two other underwing 330 liter (87 gal) tanks each result in an additional 660 liters (174gal).

There is a luggage compartment with 0,17m^3 (6ft^3) and capacity for 30kg (66lb) of cargo located behind the cockpit and accessed by the left side of the aircraft.

The cell supports a 6,000 hour life cycle with the load factor limit G of + 6/-3.0 and operation in -45° C (-49° F) to + 50°C (+ 122°F).

Chord at centre line is 2.43m (7.97ft), root chord 2.30m (7.55ft), aspect ratio of 6.40; Dihedral at 30 per cent chord is 5° 30 ‘. Flap single slotted.

Elevator deflection of 20° downward and 18° upward. Rudder deflection (right and left) is 25°. Aileron deflection of 15° downward and 17° upward. Flap maximum deflection is 35°.

Powerplant

A Pratt & Whitney PT6A-25C of 750shp power handling a Hartzell propeller constant pitch tripod HC-B3TN-3C/T10178H-8R. It is reversible to aid in braking during landing or manoeuvring on ground allowing the airplane to move backwards. Power to weight ratio (Kg/SHP) 3.4. Gas generator maximum speed of 38,100 rpm. Power turbine maximum speed of 33,000 RPM. Propeller maximum speed of 2,200 rpm.

Maximum take-off rating is 750shp at 30.5° C (86.9° F); maximum continuous rating is 750shp at 30.5° C (86.9° F); maximum climb rating of 700shp at 28.3° C (82.94° F); maximum cruise rating is 700shp at 19.4° C (66.92° F); and maximum rating for inverted flight is 750shp. On inverted flights the time limit is 30 seconds; vertical flight nose up is 15 seconds; vertical flight nose down is 15 seconds; and on zero G is 7 seconds.

Dimensions

Wingspan 11.14m (36.55ft); length 9.86m (32.35ft) – do not have probe; height 3.40m (11.15ft); wing area 19.40m 2 (208.8 sq ft); tailplane span 4.66m (15.29ft); wheelbase: 3.16m (10.37ft); wheel track 3.76m (12.34ft). Static propeller clearance is 0.33m (1.08ft).

Weights

Empty weigh 1,790kg (3,946lb); maximum ramp weight is 3,195kg (7,044lb), maximum take-off weight 3,175kg (7,000lb), maximum landing weight is 2,800kg (6,173lb).For aerobatic configuration, the maximum take-off and landing weight in 2,500kg (5,622lb).

Performance

Maximum take-off weight at sea level with 15.24 meters (50ft) of obstacle is 524 meters (1,719ft). Landing distance at maximum take-off weight at sea level with 15.24 meters (50ft) of obstacle is 590 meters (1,935ft). Rate of climb is 786 meters per minute (2,578ft per minute). Service ceiling is 8,686 meters (28,500ft). Maximum cruise speed (true airspeed, best level) is 440km/h (237knots). Range (without external fuel tanks) is 1,918km (1,035nm). Altitude of operation is from -304 meters (-1000ft) to 9,144 meters (30,000ft).

With 2,550kg (5,622lb), the diving speed is 600km/h (324knots); never exceed speed is 538km/h (291knots); stalling speed without flaps is 139km/h (75knots); stalling with flap is 127km/h (69knots). Landing gear extension speed is 227km/h (150knots).

Avionics

VHF communication system from 118 to 135Mhz; channel spacing of 50Mhz; 360 channels. The audio subsystem allows the VHF and navigation subsystem to receive audio control and intercommunication between the pilots in hot or cold microphone selection. Navigation system commitment of a gyromagnetic compass, which the magnetic heading of the plane is given by course indicators and radio magnetic indicators; an automatic digital ADF operating from 200Khz to 1,750Khz band in steps of 1Khz; two VOR/ILS/Marker Beacon and DME subsystems and to ATC transponder working on modes A or C.

Armament

All armament is conventional, non-guided and limited to 1 ton (2,204lb) divided in four underwing pylons with 250kg (550lb) capacity each.

Two external fuel tank with 330 litres (87 gal) each used on inner pylons.

Bombs: up to four Mk 81 120kg (264lb); or four Mk 82 230kg (507lb), conventional or low drag; multiple exercise bombs BEX-11 (11kg - 24lb) or Mk 76 Mod. 5 of the same weight depending of the type of the rack. Up to four incendiary bombs BINC-200 (209kg - 460lb) with stabilizer tail or two BINC-200 without stabilizer tail.

Rockets: up to eight High Velocity Aircraft Rocket of 127mm (5in); up to four LM37/7A SBAT 37mm (1.45in) with seven rockets for training purposes; up to four LM 70/7A SBAT 70mm (2.75in) or LAU-32 with seven rockets each; up to four LM70/4 SBAT 70mm (2.75in) with four rockets; EQ-LMF-70/7A 70mm (2.75in) low drag rocket pod.

Machine guns: maximum two used only at inner pilons, being: C-2 machine gun pod with one FN Herstal MAG 58P 7,62mm (.30in) with 250 rounds each; or FN Herstal Twin MAG Pod, on each pod having two machine guns MAG 58P with 500 round per pod. FN Herstal Heavy Machine Pod (HMP) with one M3P machine gun of 12.7mm (.50in) and 250 rounds; and FN Herstal Rocket Machine Gun Pod with one MP3 machine gun of 12.7mm (.50in) and 250 rounds plus three non-guided 70mm (2.75inches) rockets.

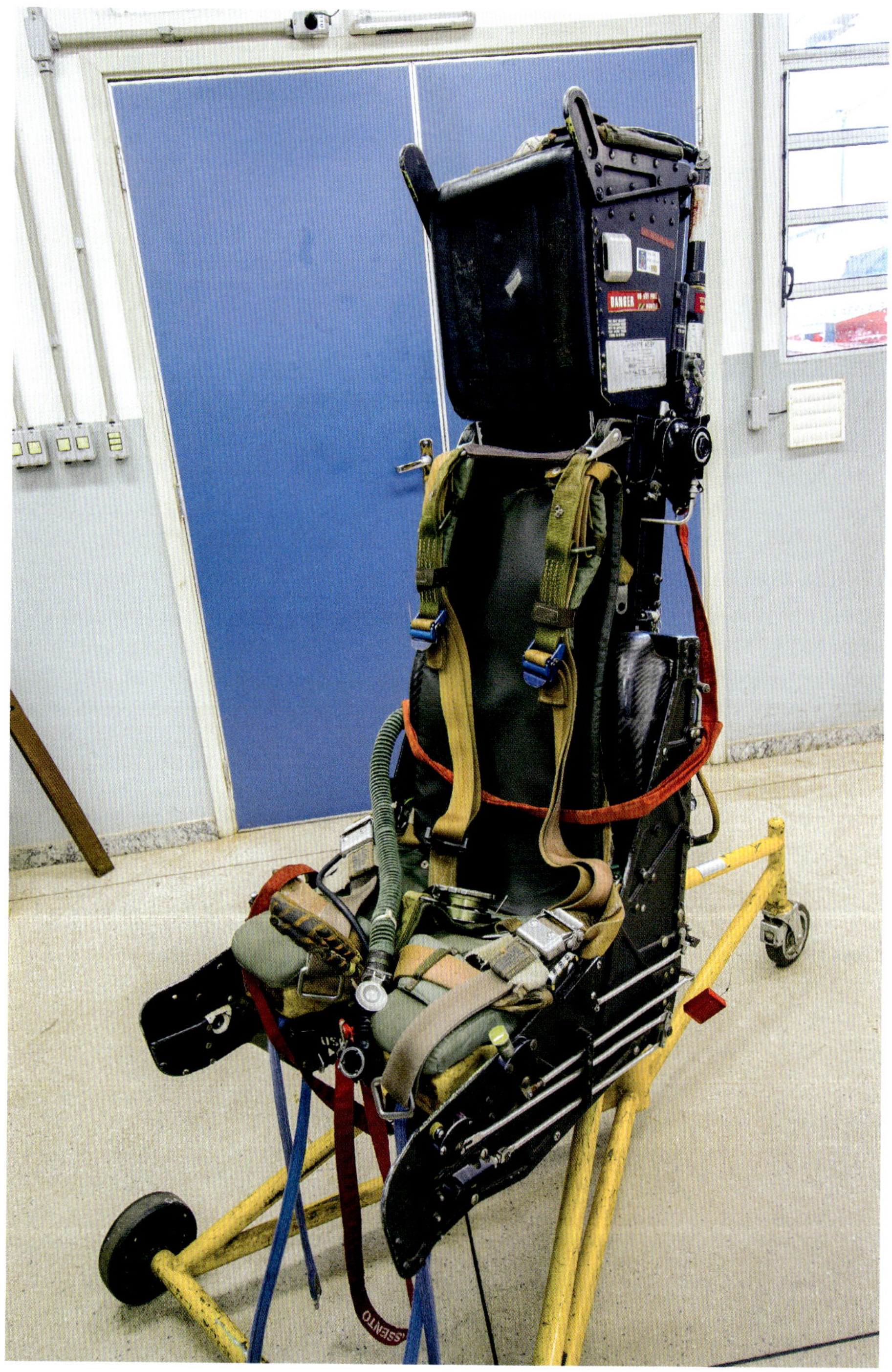

Complete Martin-BakerMk 8L ejection seat. (João Paulo Moralez)

Except for the absence of a radar scope, the cockpit architecture and instrument panel are very similar to a third-generation fighter. Seen here is the front cockpit: alarm panel, switches, ejection seat, single lever, mirrors and levers helped reduce the gap between turboprop and jet trainer. Note the green storm scope display at the lower right. (João Paulo Moralez)

The rear cockpit is almost identical to the front, but without mirrors. (João Paulo Moralez)

EMB-312 Tucano FAB 1449 of the Brazilian Air Force Academy at Pirassununga (SP) in 2016. Overall White (FS 37722) and DayGlo (FS 38915) with spinner and anti-glare panel in gloss black.

FAB
AFA

SALVAMENTO
T-27
FAB
1449

Shorts

SHORT S.312 TUCANO

Manufacturer

Embraer variant Embraer EMB-312 Tucano built under licence by Short Brothers in Belfast, Northern Ireland, to meet an order from the RAF for 130 aircraft with options for another 15. It was also sold to Kenya (12 aircraft designated T.Mk 51) and Kuwait (16 aircraft designated T.Mk 52).

The first prototype made its inaugural flight on 2 February 1986. The first production aircraft flew on 30 December of that year and was delivered to the RAF in June 1987. Total of 160 aircraft were produced.

Embraer provided 24 complete structural kits (wing, fuselage and stabilizers) for the assembly of the first aircraft; the main landing gear for all short-haul aircraft; the first prototype; 40 sets of structures and parts for canopies.

Production

Total 160 aircraft produced for the UK (130 including one prototype), Kuwait (16) and Kenya (12). One aircraft was lost in an accident, during rehearsals, before being delivered to Kenya. Another one was damaged in an IRA terrorist attack at the Short Brothers factory in Belfast, and its construction was not completed.

Role

Basic training, advanced training, fighter pilot training with weaponry, ground attack and fighter. It is aerobatic and IFR capable.

Crew

In tandem and equipped with two ejection seats Martin-BakerMk 8LC 0-70 (zero altitude and 70 knots – 130km/h). The instructor sits on a raised seat in the rear seat to observe the actions of the student sitting in the front seat. The takeoff, when soloed, can only be made with the pilot occupying the front position of the aircraft.

Analog panel with dual control, monomannel for engine control, rear view mirrors and a conventional reflex sight (Kenya and Kuwait).

Airframe and systems

Canopy divided by a central frame with its own explosive charge for destruction of the transparent part and subsequent ejection of the pilot. Without windshield. Air-conditioned, non-pressurized, 2024 aluminum alloy frame and cantilever wings and feather construction. Retractable tricycle landing gear with single tires and fully protected by doors made of composite material when collected.

The wings have two fuel tanks each for 694 liters (183 gal) in total. Two other 330 liter (87gal) tanks each result in an additional 660 liters (174 gal).

There is a luggage compartment with 0,17m^3 (6ft^3) and capacity for 30kg (66lb) of cargo located behind the cockpit and accessed by the left side of the aircraft.

The cell supports a life cycle of 12,000 hours with the load factor limit G of + 7/-3.6.

Chord at centre line is 2.43m (7.97ft), root chord 2.30m (7.55ft), aspect ratio of 6.40; dihedral at 30 per cent chord is 5°30‘. Flap single slotted. Elevator deflection of 20° downward and 18° upward. Rudder deflection (right and left) is 25°. Aileron deflection of 15° downward and 17° upward. Flap maximum deflection is 35°.

Powerplant

A Garrett TPE 331-12B of 1,110shp power handling a Hartzell four-blade constant-pitch propeller HC-B4TN-5S/T9327K. It is reversible to aid in braking during landing or manoeuvring on ground allowing the airplane to move backwards. Propeller maximum speed of 2,000rpm. Power to weight ratio is 2.4 (kg/SHP).

Maximum take-off rating is 1,100shp at 30.5° C (86.9° F); maximum continuous rating is 1,100shp at 30.5° C (86.9° F). On inverted flights the time limit is 30 seconds; vertical flight nose up is 15 seconds; vertical flight nose down is 15 seconds; and on zero G is 15 seconds.

Dimensions

Wingspan 11.28m (37ft); Length 9.85m (32.31ft) – do not have probe; height 3.40m (11.15ft); wing area 19.3m 2 (208 sq ft); tailplane span 4.66m (15.29ft); wheelbase: 3.12m (10.24ft); wheel track 3.76m (12.34ft). Static propeller clearance is 0.32m (1.04ft).

Weights

Empty weigh 2,017kg (4,447lb); maximum ramp weight is 3,300kg (7,275lb); maximum take-off weight 3,280kg (7,231lb), the maximum landing weight is 2,650kg (5,842lb). For aerobatic configuration, the maximum take-off and landing weight in 2,650kg (5,842lb).

Performance

Maximum take-off weight at sea level with 15.24 meters (50ft) of obstacle is 500 meters (1,640ft). Landing distance at maximum take-off weight at sea level with 15.24 meters (50ft) of obstacle is 590 meters (1,935ft). Rate of climb is 1,082 meters per minute (3,550ft per minute). Service ceiling is 10,363 meters (34,000ft). Maximum cruise speed (true airspeed, best level) is 494km/h (267knots). Range (without external fuel tanks) is 1,665km (900nm).

Never exceed speed is 518km/h (279knots); stall speed with flaps and gear down is 128km/h (69knots). Landing gear extension speed is 227km/h (150knots).

Avionics

TACAN antennas; ILS capability; ATC transponder; V/UHF communication systems; VOR, DME and Marker Beacon.

Armament

All armament is conventional, non-guided. RAF variant does not have a provision for armaments. Kenya and Kuwait planes have four underwing stations limited to 454kg (1,000lb) total.

Bombs: up to four Mk 81 120kg (264lb); two Mk 82 230kg (507lb); multiple (eight) exercise bombs BEX-11 (11kg - 24lb) or Mk 76 Mod. 5 (11kg - 24lb) with smoke cartridges on CBLS 100 pod.

Rockets: up to four LAU-32 with seven rockets each.

Machine guns: FH Herstal Ttwin MAG Pod, on each pod having two machine guns MAG 58P with 1.000 round per pod. FN Herstal Heavy Machine Pod (HMP) with one M3P machine gun of 12.7mm (.50in) and 250 rounds; and FN Herstal Rocket Machine Gun Pod with one M3P machine gun of 12.7mm (.50in) and 250 rounds plus three non-guided 70mm (2.75in) rockets.

Two external fuel tank with 330 litres (87gal) each used on inner pylons.

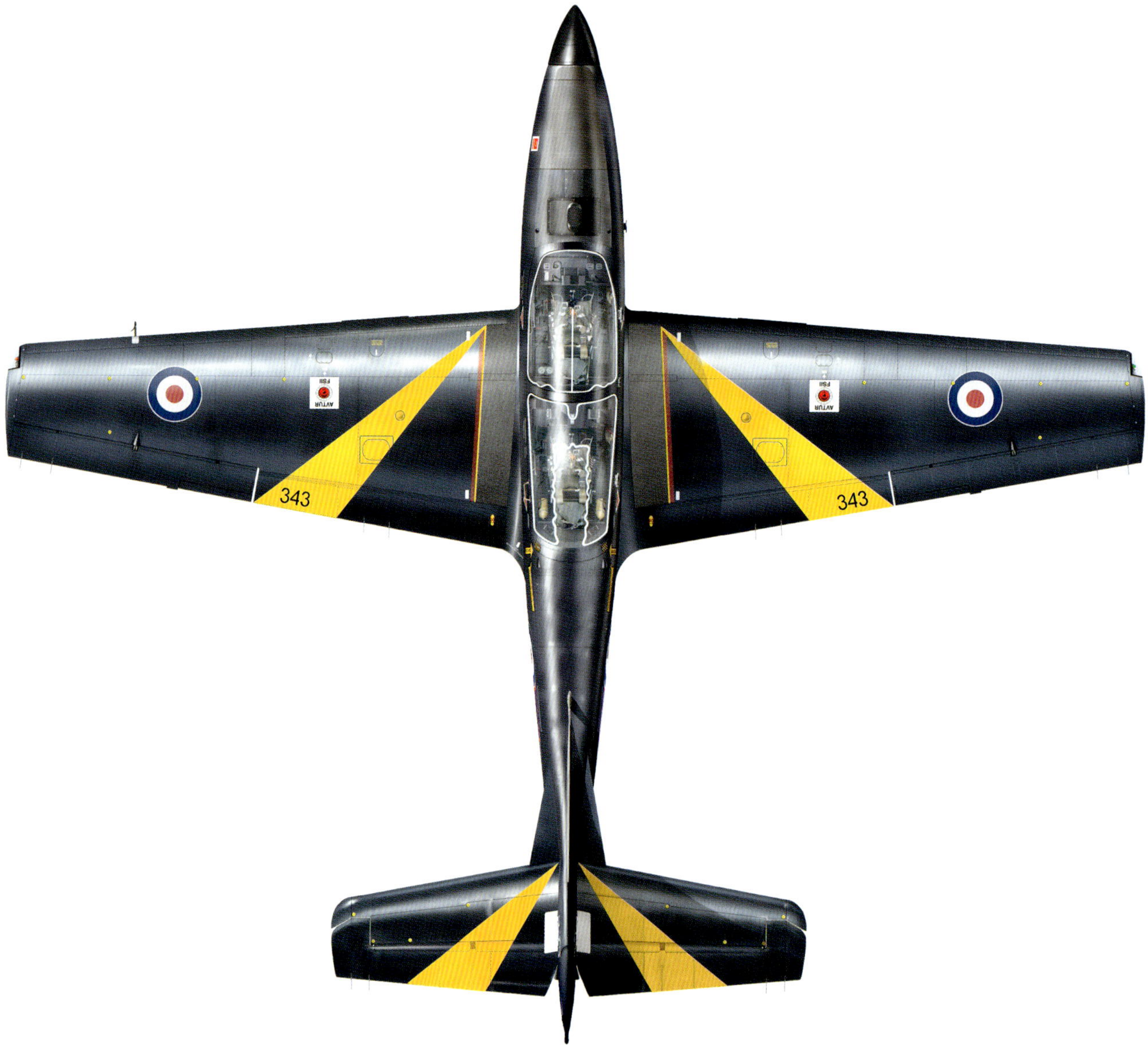

Short Tucano T.Mk 1 ZF343 from No 1 Flying Training School, No 72 (Reserve) Squadron at RAF Linton-On-Ouse in 2015. Overall Gloss Black (FS 27040) with the spinner in black and white.

343
ZF343
343

APPENDIX II: ARTWORKS

Embraer pre-studies

The EMB-301 was Embraer's first study for the Brazilian Air Force's next trainer. It featured tandem seating, four hardpoints under the wings and a 380hp Lycoming TIO-541-E1C4 piston engine with a three-blade variable pitch propeller. Embraer was not satisfied, judging it too simple for an advanced trainer. May 1977.

The EMB-311 was the response to Embraer's own criticism of the EMB-301. Fitted with a 750shp Pratt & Whitney Canada PT6 turboprop, it had improved manoeuvrability, speed, six underwing hardpoints and much greater fuel capacity. It still retained the basic shape of the T-25 and was now fitted with ejection seats. The air force gave more attention to this design study, discarding the EMB-301. May 1977.

The very first configuration for the EMB-312. The removal of the ejection seats provided a 150mm (5.9in) reduction in fuselage length, but the second seat was higher than the front seat, providing the instructor with an improved view of the student. The fuselage still bears a similarity to the T-25. August 1977.

Embraer prototypes

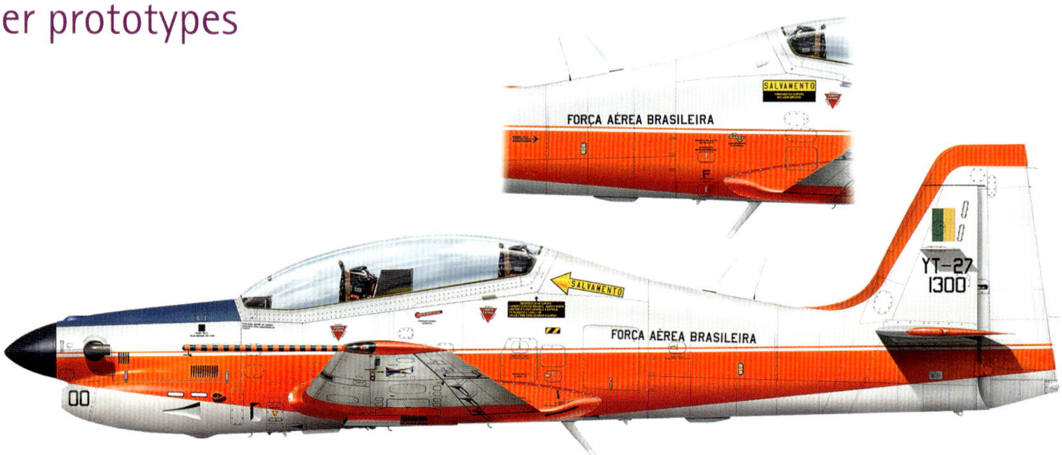

The eighth configuration that led to the first flying prototype was totally different from the initial studies. Designated YT-27 by the air force (Y = prototype, T = trainer), it was first flown on 16 August 1980. It received the registration 1300, and was lost after a forced landing on 9 April 1984. It had flown 786 hours. Painted gloss White (FS 37722) and DayGlo (FS 38915). Anti-glare panel is flat Blue Angels Blue (FS 15050).

EMB-312 Tucano

Together with PP-ZTT, PT-ZTK was one of the various examples used by Embraer for test campaigns, airshows or demonstration flights to the press and possible clients. It was sold to Angola in 2003, after a major overhaul at São José dos Campos. Gloss Green (FS 14090) with gloss White (FS 37722).

Shorts Tucano

EMB-312 PP-ZTC was a proof-of-concept airframe for the Short Tucano programme for the UK. The construction number was 312.007, coincidentally recalling fiction secret agent James Bond (007). It received a 1,100hp four-blade Garrett TPE-331-12B and, except for the nose, retained all the original structure including canopy. It was designated EMB-312G2 Tucano and was first flown on 14 February 1986 by Major Gilberto Pedrosa Schittini and engineer Luiz Fernando Oliveira. Once in UK it received the test registration G-14-007 and then the British registration G-BTUC ('British Tucano'). It was withdrawn from service on 20 December 1996 and preserved at Ulster Aviation Society Heritage Centre Langford Lodge. It wore Red (FS 31302) and gloss White (FS 37722) and is shown as it appeared before its first flight.

Brazilian Air Force

FAB 1329, 1°/3° GAV 'Esquadrão Escorpião', Boa Vista Air Base, armed with low-drag BAFG-120 (230kg/507lb) bombs. The aircraft took part in actions against FARC guerrillas during Operation Green Carpet in 1999. Colours were Dark Grey (FS 25104) or Grey (FS 36118), depending on the aircraft. Lower surfaces were Grey (FS 36622).

FAB 1438, another EMB-312 from 1°/3° GAV 'Esquadrão Escorpião' with the Southeast Asia colour scheme, in 2001. Although more suitable for the Amazon region, the air superiority scheme was more common among the Amazon squadrons. Colours were Dark Tan (FS 30219), Dark Green (FS 34079), Medium Green (FS 34102) and Grey (FS 36622).

FAB 1447 from 1°/3° GAV 'Esquadrão Escorpião' in 2002. The most common armament for the Brazilian EMB-312 was the seven-tube Avibras 70/7M5A pod for 70mm (2.75in) unguided rockets, C-2 machine gun pods with a 7.62mm (0.3in) FN MAG and BAFG-120 (120kg/264lb) bombs. T-27 FAB 1447 has four BAFG-230 (230kg/507lb) bombs. A mixed weapons configuration was uncommon. Colours were Dark Grey (FS 25104) or Grey (FS 36118), depending on the aircraft. Lower surfaces were Grey (FS 36622).

FAB 1448 in 1999. This is one of only 15 FAB EMB-312s capable of employing heavier weapons like the FN Herstal HMP 12.7mm (0.5in) machine gun, although it was never used by the FAB on the Tucano. The aircraft is armed with a 200kg (409lb) BINC-200 napalm tank. 2°/3° GAV 'Esquadrão Grifo' was based at Porto Velho to cover a strategic region of the Amazon border. The unit operated the EMB-312 for air defence, escort, attack, close air support, interception and border patrol. The 'PV' code indicates that FAB 1448 belongs to Porto Velho. On the right side is the emblem of III Air Force. Colours were Dark Grey (FS 25104) or Grey (FS 36118), depending on the aircraft. Lower surfaces were Grey (FS 36622).

Very well used and exposed to weather, EMB-312 FAB 1348 previously belonged to 1°/14° GAV 'Esquadrão Pampa'. In 2006 it was stationed at Campo Grande Air Base with 3°/3° GAV 'Esquadrão Flecha'. The unit flew air defence, escort, attack, close air support, interception and border patrol missions in Brazil's mid-west. The logo on the tail refers to the centenary of Santos-Dumont's flight in the 14-bis, on 23 October 1906. Colours were Dark Tan (FS 30219), Dark Green (FS 34079), Medium Green (FS 34102) and Grey (FS 36622).

FAB-1344 in 2001, with two low-drag EQ-LMF 70/7 rocket pods. 1°/5° GAV 'Esquadrão Rumba' became the unit responsible for the formation of new attack pilots. Based at Natal Air Base, tail-coded 'NT', the unit used EMB-312s in Southeast Asia camouflage and air superiority two-tone grey. The tone from above varies from Dark Grey (FS 25104) to Grey (FS 36118), depending on the aircraft. Lower surfaces were Grey (FS 36622).

The front-line squadrons of the Brazilian Air Force received some EMB-312s for liaison, instrument and navigation training and interception of light aircraft trafficking drugs and weapons. Seen as it appeared in 2000, FAB 1381 belonged to 1°/14° GAV 'Esquadrão Pampa', Canoas Air Base, at the beginning of the 1990s. Today the unit operates the modernised Northrop F-5EM/FM. Camouflage is Southeast Asia with Dark Tan (FS 30219), Dark Green (FS 34079), Medium Green (FS 34102) and Grey (FS 36622). This aircraft served with the unit from 4 August 1996 to 4 July 2002. The 'CO' tail code indicates Canoas, according to the ICAO airport designation.

The 2ª Esquadrilha de Ligação e Observação (2ª ELO, 2nd Liaison and Observation Flight) was the first unit in Brazil to use EMB-312 as a weapons platform. They developed doctrine and employed a wide range of bombs, rockets and machine guns. The aircraft received the pilots' names: FAB 1414 was the personal EMB-312 of the commander, Captain Pereira Pinto, in 1987. At the rear seat is the name of his mechanic, Sergeant Prado. The squadron remained at São Pedro da Aldeia Naval Air Base, in Rio de Janeiro, until 1995. Southeast Asia camouflage with Dark Tan (FS 30219), Dark Green (FS 34079), Medium Green (FS 34102) and Grey (FS 36622).

Third and final front-line fighter squadron to receive the EMB-312 for support missions was the 1° Grupo de Defesa Aérea (1° GDA, 1st Air Defence Group) at Anápolis. Home of the Dassault Mirage IIIE/D, Mirage 2000C/B and now the Northrop F-5EM/FM, the group also provided air-to-air instruction for the Boa Vista and Porto Velho Tucano squadrons. FAB 1437 is shown as it appeared in 1995. The 'AN' tail code is for Anápolis and the jaguar on the top of rudder is the symbol of the group. The group also flew some EMB-312s in air superiority colours. For this aircraft, colours were Dark Tan (FS 30219), Dark Green (FS 34079), Medium Green (FS 34102) and Grey (FS 36622).

In the early 1990s the 1° Grupo de Aviação de Caça (1° GAVA, 1st Fighter Aviation Group), based at Rio de Janeiro and equipped with the Northrop F-5E (now F-5EM/FM), received some EMB-312s for liaison, IFR training, interception of light aircraft trafficking drugs and weapons employment. The group has two squadrons: 'Senta a Púa' and 'Rompe Mato'. Like the fighters, the EMB-312 was used in a pool by both squadrons, serving the two squadrons when needed. FAB 1377 is shown as it appeared in the late 1990s with two C-2 machine gun pods with a 7.62mm (0.3in) FN MAG 58P. Uniquely, the light grey was different to the rest the fleet, apparently Medium Grey (FS 36176). Nose and rudder were Grey (FS 36622). Darkest upper colour was Dark Grey (FS 25104).

FAB 1303 was the first production EMB-312, delivered to the Academia da Força Aérea (AFA, Air Force Academy), Pirassununga, on 29 September 1983, where it served with the 1° Esquadrão de Instrução Aérea (1° EIA, 1st Air Instruction Squadron). Gloss White (FS 37722) and DayGlo (FS 38915) with spinner and anti-glare panel in black. FAB 1303 has been withdrawn from service.

EMBRAER
Defesa & Segurança

Celebrating 30 years of EMB-312 operations in Brazil, logged on 29 September 2013, the Air Force Academy painted FAB 1361 in this attractive scheme colour, including a toucan in flight applied on the fuselage. This aircraft has limitations for high-G flying and is only used for instruction flights with no aerobatic manoeuvres. It is expected to be withdrawn and preserved at the Air Force Academy.

FAB 1311 was the personal mount of Lieutenant Colonel Aviator Soares Filho, the second commander of Esquadrilha da Fumaça (Smoke Squadron) during the Tucano era, from 1986 to 1989. Gloss Red (FS 31302), used from 1983 until the beginning of 2000s, was inspired by the Northrop F-20 Tigershark colour scheme. The name on the rear seat is Captain Hélio Magajeviski.

On 25 February 2002, after a two-year break in operations to change the wings of its fleet, the Smoke Squadron returned with these blue, green and yellow colours, representing a more 'nationalist' scheme. FAB 1371 was the aircraft of Lieutenant Colonel Aviator Wagner de Almeida Esteves, the unit's commander from 2011 to 2012. The flags represent the countries visited by the team and the blue logo in the centre of the fuselage commemorates the 60th anniversary of the team, celebrated on 14 May 2012. Dark Blue (FS 15056) overall.

At the same time as the Air Force Academy and Smoke Squadron received their first EMB-312s, the Pesquisa e Ensaios em Voo (PEV, In-Flight Research and Testing), an air unit of Centro Técnico Aeroespacial (CTA, Aerospace Technical Centre) received the third prototype Tucano. YT-27 1302 was used for a wide range of tests including weapons platform validation. It dropped Mk 76 bombs (11kg/24lb); SBAT 37 rockets (37mm/1.45in); SBAT 70 rockets (70mm/2.75in); C-2 type pods, each equipped with a FN MAG 58P 7.62mm (0.3in) machine gun; Avibras M2A1 napalm bombs (180kg/400lb); AN-M57 bombs (104kg/230lb); and the BLG-120 cluster bomb, in a configuration seen in 2004. Colours are gloss White (FS 37722) and DayGlo (FS 38915). Anti-glare panel in flat black.

For fewer than 15 days in mid-1986, FAB 1383 received a shark mouth on the nose for participation in a display alongside two other EMB-312s, during a presentation for the authorities at the Parque de Material Aeronáutico de Lagoa Santa (PAMA-LS, Lagoa Santa Aeronautical Material Park). The shark mouth was painted using masks by Sub-official Raul Sabino and was removed after the presentation. Southeast Asia camouflage with Dark Tan (FS 30219), Dark Green (FS 34079), Medium Green (FS 34102) and Grey (FS 36622).

Foreign operators

The FAN's EMB-312 was very active during the 1990s civil war. Almost all operations were carried out in a two- or four-ship formation, each aircraft using four seven-tube LAU-32 pods for SBAT 70 rockets (70mm/2.75in). The batch of six aircraft acquired new from Embraer did not have the shark mouths. Only after Angola bought six second-hand airframes from Peru did it start using this camouflage pattern and the shark mouth. R-461, ex-Peruvian FAP 427, was carrying LAU-32 rocket pods at Luanda in 2010. Upper colours were Light Earth (FS 30257), Intermediate Green (FS 14109) and Medium Green (FS 14115). Underside colour was Grey (FS 36622).

The first 15 EMB-312s for the Fuerza Aérea Argentina came from Brazilian Air Force stocks – at that time, between 25 and 30 airframes were stocked at Lagoa Santa. To meet Argentina's needs, they retained the Southeast Asia camouflage, adding only DayGlo (FS 38915) on the wingtips and tail. All were assigned to the Escuela de Aviación Militar (EAM, Military Aviation School), at Córdoba, in the mid-1980s. E-128 wears a logo marking the centennial of military aviation in Argentina, celebrated in 2012. Colours are Dark Tan (FS 30219), Dark Green (FS 34079), Medium Green (FS 34102) and Grey (FS 36622).

E-110 of the Military Aviation School. The Fuerza Aérea Argentina's EMB-312s changed to an air superiority scheme at the beginning of 2015. Although Argentina has pylons for the Tucano fleet, the aircraft has never flown with weapons fitted. Colours are Dark Sea Grey (FS 36099) and Light Ghost Grey (FS 36375) with DayGlo (FS 38915) rudder and wingtips. E-114, E-119 and E-129 were painted in the same scheme.

FAC 2254, seen as it appeared in 2004, belonged to Grupo de Combate 21 (Combat Group 21) and is armed with four Mk 82 bombs for attack missions. Initial camouflage for the Colombian EMB-312 was Grey (FS 36622) with yellow, blue and red reflecting the national colours. Anti-glare panel in flat black and spinner with a spiral design in black.

Most Colombian EMB-312s wore individual markings and colours. FAC 2253 from the Escuadrón de Combate 212 (Combat Squadron 212), of Grupo Aéreo 21 (Air Group 21), Comando Aéreo de Combate No. 2 (Combat Air Command 2) at Apiay, is seen as it appeared in 2008. Grey Medium (FS 36270) forward fuselage, Dark Gull Grey (FS 36231) from the middle of fuselage of the tail, and Grey (FS 36622). Anti-glare panel in flat black and spinner with a spiral design in white.

Exotic EMB-312 FAC 2256, from Escuadrón de Combate 212, received a shark mouth, plus a sticker reading 'We're the Force' and a special scheme that was different on each side, to celebrate 80,000 flight hours by the Tucano in Colombia in 2013. On the right, a toucan is pictured riding a bomb. The aircraft belonged to 2nd Lieutenant Yate 'Cron' and Lieutenant Guarin 'Guahibo'. Colours were Dark Gull Grey (FS 36231) and Grey (FS 36622). Anti-glare panel in flat black and spinner with a spiral design in white.

Egypt became the third largest operator of the EMB-312, with 54 airframes in total, almost all of them built locally under licence. In the mid-1980s they wore a paint scheme similar to that of the FAB's Air Force Academy with gloss White (FS 37722) and DayGlo (FS 38915) and a flat black anti-glare panel, as seen on EMB-312 '518'. This belonged to the 101st Air Brigade at Bilbeis.
Note: all EMB-312s produced in Egypt have a red anti-collision light on top of the tail.

After 2000, Egypt adopted this new scheme for its Tucano fleet, standardising it with the air force's other trainers: gloss White (FS 37722) overall and stripes, bands and details in Insignia Blue (FS 35044). In 2001, EMB-312 '544' was assigned to the 201st Air Brigade at Bilbeis.

After 2000, this Grey (FS 36118) camouflage was applied to the Armée de l'Air EMB-312F fleet to increase aircraft conspicuousness. Assigned to the École de l'Air, at Salon de Provence, serial number 462 '312-JG' is see as it appeared in 2007, when assigned to the Centre de Formation Aérienne Militaire Initial (Initial Air Military Formation Centre). The spinner was polished metal. The logo of Division des Vols 05.312 appears on the tailfin.

EMB-312F serial number 487 '312-UF' as it appeared in 2007, with the high-conspicuousness Grey (FS 36118) camouflage.

After the start of operations in France, the EMB-312F received this gloss White (FS 37722) and DayGlo (FS 38905) scheme. Serial number 459 '312-JD' is seen as it appeared in 2009. It served with the École de l'Air at Salon de Provence. Before being withdrawn from service in July 2009, it received an artwork of a pilot on the tailfin and the inscription 'Bye'.

First export client for the EMB-312 was Honduras, which took 12 examples in 1984 and 1986. All were used for advanced training and conversion to jet aircraft, at the Academia Militar de Aviación (Military Aviation Academy), Colonel José Enrique Soto Cano Air Base, located in Palmerola. For COIN missions, Honduras received Mk 81 and Mk 82 bombs, Avibras 70/7M5A pods for unguided 70mm (2.75in) rockets and FN Herstal Twin MAG Pods with two 7.62mm (0.3in) MAG 58P machine guns. FAH 256, seen as it appeared in 2013, had a heavily weathered paint scheme of Marine Green (FS 34052), Green (FS 34138), Rusty Red (FS 30252) and Field Drab (FS 33105).

'Aggressive' Honduran EMB-312 FAH-251, seen as it appeared in 2016, received a Marine Green (FS 34052) and Grey (FS 36622) scheme. The new paint did not properly cover the old camouflage, which was visible in some lighting conditions. The ejection seat triangles were non-standard and the tailfin emblem is for the 1° Escuadrón de Entrenamiento (1st Training Squadron). The aircraft is armed with unguided 70mm (2.75in) rockets and two 11kg (24lb) Mk 76 bombs.

The Islamic Revolutionary Guard of Corps Air and Space Force EMB-312 fleet is dedicated to COIN missions employing two underwing seven-tube pods for unguided 70mm (2.75in) rockets. Camouflage is similar to that of the Islamic Republic of Iran Air Force, with Brown Special (FS 30140), Tan Special (FS 30400), Dark Green (FS 34079) and Grey (FS 36622) and a flat black anti-glare panel. Serial number 15-2413 is seen as it appeared in 2014, with the logo of the Islamic Revolutionary Guard of Corps Air and Space Force on the tail.

The first 10 airframes for Iraq were assembled in Brazil with Egyptian assistance, while the remainder were made in Egypt. Those made in Brazil were gloss Red (FS 31302) overall with white and black stripes and details. EMB-312 '6707' served with the Iraqi Air Force Academy at Tikrit in the mid-1980s.

Iraq adopted the same colour scheme used by Egypt and Brazil for its training aircraft: gloss White (FS 37722) and DayGlo (FS 38915) with a flat black anti-glare panel. This aircraft, '6743', was also assigned to the Air Force Academy in Tikrit and these colours were used until 2003, when the aircraft was withdrawn from service.

Using the same training colours as the UK, Kenya received its Short Tucanos in Red (FS 31302) and White (FS 37875), with a white spinner and black anti-glare panel. Wings also had Light Gull Grey (FS 36440). Designated T.Mk 51, KAF 811 is seen as it appeared in 2000, at the Kenya Air Force Flying Training School, Moi Air Base, Mombasa International Airport.

Half of 12 Kenya's Tucano T.Mk 51s received this 'operational' camouflage of Dark Green (FS 34079), Dark Sea Grey (FS 36173) and Grey (FS 36622). Black anti-glare panel and Dark Green spinner. Based at the Kenya Air Force Flying Training School, at Moi, in the suburbs of Nairobi. All can be fitted with Mk 81 and Mk 82 family bombs, seven-round 70mm (2.75in) unguided rocket pods and FN Herstal Heavy Machine Pods (HMP) with 12.7mm (0.5in) M3P machine guns.

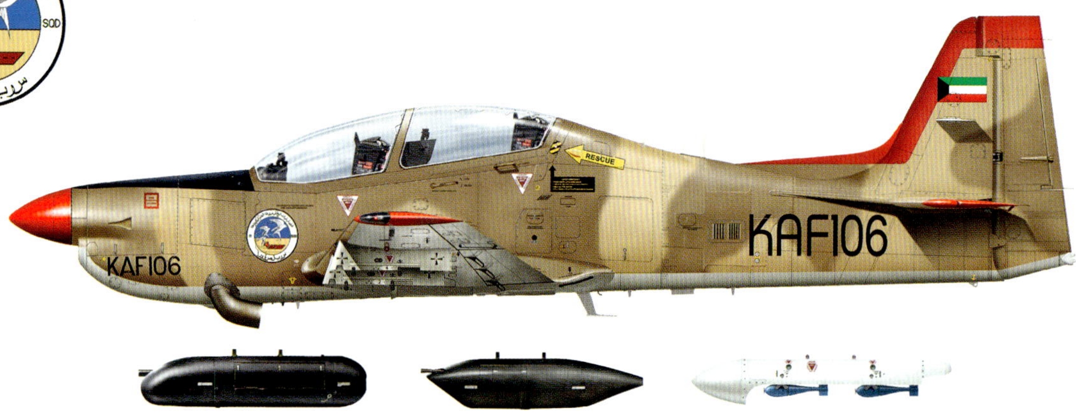

The Kuwait Air Force operated a fleet of 16 Short Tucano T.Mk 52s assigned to the 19th Flight Training Squadron at Ali Al Salem Air Base. All have been used for advanced training since their delivery in 1995, but they could be armed with Mk 81 250lb (119kg); up to eight BEX-11 exercise bombs of 11kg (24lb), or Mk 76 Mod 5 bombs of the same weight, in a CBLS 100 pod; 70mm (2.75in) unguided rocket pods; FN Herstal Twin MAG Pods, each pod having two MAG 58P machine guns with 500 round; and FN Herstal Rocket Machine Gun Pods with one M3P 12.7mm (0.5in) machine gun and 250 rounds plus three unguided 70mm rockets. Colours were Dark Earth (FS 30118), Middle Stone (FS 30266) and Grey (FS 36622). Spinner, wingtips and elevators in DayGlo (FS 38915).

Mauritania became an operator of the EMB-312F after France retired its fleet. 5T-MAT/499 is seen as it appeared in 2011, armed with the FN Herstal Heavy Machine Pod (HMP) with a 12.7mm (0.5in) M3P machine gun. The aircraft belongs to the École Militaire de l'Air, at Base Aérienne 210 Atar. Mauritania is the only country to adopt desert camouflage for the Tucano, with a wraparound Light Sand (FS 32563) and Light Earth (FS 30257).

Armed with the Avibras 70/7M5A pod for unguided 70mm (2.75in) rockets and an FN Herstal Twin MAG Pod with two 7.62mm (0.3in) MAG 58P machine guns, EMB-312 FAP 1052 belongs to Tercero Escuadrón de Caza 'Moros' (3rd Fighter Squadron), subordinated to Grupo Aerotáctico (Aerotactic Group) at Ñu Guazú Air Base in Luque, near the capital Asunción. Colours are Southeast Asia with Dark Tan (FS 30219), Dark Green (FS 34079), Medium Green (FS 34102) and Grey (FS 36622)

The Fuerza Aérea del Peru received its first batch of 20 EMB-312s from 1987 onwards for advanced training. They equipped Escuadrón Aéreo No. 512 (512 Squadron) of Grupo No. 51 (51st Group), first at Las Palmas Air Base and later at Pisco Air Base. FAP 432 is shown painted in semi-gloss White (FS 37722), DayGlo (FS 38915) and with an anti-glare panel in flat Blue Angels Blue (FS 15050). A GPS is installed above the cockpit instrument panel, left of the gunsight. The logo on the tail celebrates the centennial of José Abelardo Quiñones Gonzales, national hero and patron of the Peruvian Air Force, in 2014. At the top of the rudder is the logo of Servicio de Mantenimiento de la Fuerza Aérea del Perú and below it, the logo of Grupo No. 51.

Recognising its quality as a combat platform, and after tests were conducted in the country with support from Embraer, Peru bought another 10 EMB-312s for combat purposes and interception of light drug-trafficking aircraft. All were delivered from 1991 with a camouflage of Dark Green (FS 34079), Green (FS 34102) and Dark Tan (FS 30219), and a shark mouth on the nose. The belly was Grey (FS 36622). In the mid-1990s Peru introduced NVGs and this camouflage pattern created confusion when seen through the goggles. Instead, Peru painted the aircraft in overall Dark Green, as seen here on FAP 433. It is armed with an FN Herstal Rocket Machine Gun Pod with a 12.7mm (0.5in) M3P machine gun with 250 rounds plus three 70mm (2.75in) unguided rockets. The operator is Escuadrón Aéreo No. 514 (No. 514 Squadron), at Pisco.

In 2017 Fuerza Aérea del Peru presented EMB-312 FAP 435 in low-visibility colours. Belonging to Grupo Aéreo No. 51 and based at Pisco, it is painted in overall Light Ghost Grey (FS 36375) with details in Dark Sea Grey (FS 36099). At the top of the tail is the emblem of the Servicio de Mantenimiento de la Fuerza Aérea del Perú. Above this are emblems of both Escuadrón Aéreo No. 512 and Escuadrón Aéreo No. 514.

ZF295, a Tucano T.Mk 1 from No. 1 Flying Training School, No. 72 (Reserve) Squadron at Linton-On-Ouse as it appeared in 2007. It wears overall gloss Black (FS 27040) with a spinner in black and white. The pilot was Flight Lieutenant Bobby Moore. The British flag was applied on the ventral airbrake.

No. 1 Flying Training School, No. 72 (Reserve) Squadron at Linton-On-Ouse painted Tucano T.Mk 1 ZF448 to celebrate the 90th anniversary of No. 72 Squadron. Seen as it appeared in 2007, it received the name Basuto (now Lesotho), the former British colony that donated more than 20 aircraft to the UK during World War II, including one Spitfire. Overall gloss Black (FS 27040) with spinner in black and white.

Some RAF Tucanos have been painted to represent Spitfires in World War II. Tucano T.Mk 1 ZF317 of No. 1 Flying Training School, No. 72 (Reserve) Squadron at Linton-On-Ouse received this Battle of Britain scheme. It represents the Spitfire flown by Group Captain Brian Kingcombe who received the Distinguished Service Order, Distinguished Flying Cross and Medal Bar. The ace flew with No. 92 Squadron in 1940. Carrying the code 'Q-JF', ZF317 is seen as it appeared in 2010. It was used for the RAF Tucano solo display flown by Flight Lieutenant Tom Bould and exhibition director Flight Lieutenant Martin Wintermeyer. Colours were Dark Green (FS 34079), Dark Earth (FS 30118) and Sky (FS 34424).

To celebrate the 70th anniversary of No. 72 Squadron, No. 1 Flying Training School, No. 72 (Reserve) Squadron at Linton-On-Ouse introduced this North African World War II scheme on its display Tucano T.Mk 1 in 2013. The aircraft was flown by Flight Lieutenant Andrew Fyvie and exhibition director Flight Lieutenant Dave Kirby. ZF239 was painted Dark Earth (FS 30118), Middle Stone (FS 30266) and Azure Blue (FS 35231). The spinner was red.

In 2013, No. 1 Flying Training School, No. 72 (Reserve) Squadron's Tucano T.Mk 1 solo display used ZF264, which received a special scheme for the RAF Association. Again, the aircraft was flown by Flight Lieutenant Andrew Fyvie and exhibition director Flight Lieutenant Dave Kirby. Overall gloss Black (FS 27040) with the spinner in black and white.

To mark 100 years since the start of World War I, No. 1 Flying Training School, No. 72 (Reserve) Squadron at Linton-On-Ouse painted Tucano T.Mk 1 ZF140 in this scheme for 2014. It includes poppies and the motto 'Lest We Forget'. The pilot was Flight Lieutenant Dave Kirby and exhibition director was Flight Lieutenant Jon Trueman. Overall gloss Black (FS 27040), spinner in red and black.

No. 1 Flying Training School, No. 76 (Reserve) Squadron at Linton-On-Ouse prepared ZF376 with this Tucano display scheme in 2011. The pilot was Flight Lieutenant Lt Dan Hayes and the exhibition director was Flight Lieutenant Nathan Dales. Silver (FS 17178) and Insignia Blue (FS 35044) including spinner. Black anti-glare panel.

To celebrate the Diamond Jubilee of Queen Elizabeth II (1952 to 2012), No. 1 Flying Training School, No. 76 (Reserve) Squadron at Linton-On-Ouse painted Tucano T.Mk 1 ZF269 in Insignia Blue (FS 35044), Red (FS 31302) and White (FS 37875). The pilot for the RAF Tucano display was Flight Lieutenant Jon Bond and exhibition directors were Flight Lieutenants Rich Keeling and Rob Brennan. Red spinner and black anti-glare panel.

No. 1 Flying Training School, No. 207 (Reserve) Squadron at Linton-On-Ouse prepared Tucano T.Mk 1 ZF317 with a special shark scheme for the 2009 RAF Tucano display, as presented during the Royal International Air Tattoo at Fairford. Overall gloss Black (FS 27040) with black and white spinner.

In 1992, No. 3 Flying Training School's Cranwell Poachers aerobatic team painted Tucano T.Mk 1 ZF409 in Red (FS 31302), White (FS 37875) and Blue (FS 25488) with edges in Dark Blue (FS 15090). Black anti-glare panel and spinner in black and white. The aircraft is seen as it appeared in1993. Pilot was Flight Lieutenant Karim Sachdedina.

Tucano T.Mk 1 ZF413 from No. 6 Flying Training School at Finningley, seen as it appeared in 1994. This was the first training scheme used by the RAF Tucano in the late 1980s. Colours were Red (FS 31302) and White (FS 37875). Wings also had areas of Light Gull Grey (FS 36440). Black anti-glare panel and black and white spinner.

No. 7 Flying Training School at Church Fenton created this display scheme for Farnborough 1990. Piloted by Flight Lieutenant Steve Chick, Tucano T.Mk 1 ZF145 received a toucan on the tail and an Insignia Blue (FS 35044) fuselage stripe over the traditional Red (FS 31302) and White (FS 37875) training colours. Wings also had areas of Light Gull Grey (FS 36440). Black anti-glare panel and black and white spinner.

In 1993 the Central Flying School at Linton-On-Ouse presented this RAF Tucano display scheme. Tucano T.Mk 1 ZF406 was piloted by Flight Lieutenant Neil Jackson and wore overall Oxford Blue (FS 15056). Black anti-glare panel and black and white spinner.

Celebrating the RAF's 90th anniversary, the Central Flying School at Linton-On-Ouse painted Tucano T.Mk 1 ZF210 with this display scheme in 2008. The pilot was Flight Lieutenant Stew Campbell. The aircraft was painted in gloss Black (FS 27040) with black and white spinner. The traditional RAF roundel was applied on the belly.

Private military training companies bought most of the French EMB-312F and British Short Tucanos exported to the US. The aim is to provide advanced training for US armed forces and other air forces around the world at low cost. Ex-Armée de l'Air serial number 484 now belongs to Military Tech Inc. and wears Grey (FS 36118) and Blue (FS 25622) camouflage. The purpose of the underwing pods is unknown.

RSW Aviation based in Phoenix, Arizona, specialises in advanced recovery training with the Short Tucano. It is also converting a Short Tucano with wing and structural modifications to carry laser-guided Advanced Precision Kill Weapon System (APKWS) rockets, AGM-114 Hellfire missiles, FN Herstal Heavy Machine Pod (HMP) with a 12.7mm (0.5in) M3P machine gun or FN Herstal rocket/machine gun pods with a 12.7mm M3P machine gun with 250 rounds plus three 70mm (2.75in) unguided rockets. It is also possible to install an L3/Wescam MX-15 infrared camera. Colour for the former ZF241, now N835RS, is overall gloss Black (FS 27040).

One of the many former RAF Tucanos imported to the US and sold to private owners was ZF206. It crashed on 22 June 2015, killing the renowned composer James Horner, responsible for the soundtracks to Hollywood blockbusters Titanic and Avatar. Overall gloss Black (FS 27040).

Venezuela received 31 EMB-312s in total, for different types of missions. FAV 8247 is shown as it appeared in 2013 armed with the LAU-32 pod with seven 2.75in (70mm) rockets, plus a pylon with two 11kg (24lb) Mk 76 inert exercise bombs. Assigned to Escuadrón de Entrenamiento Táctico (Tactical Training Squadron) No. 143, of Grupo de Entrenamiento Aéreo (Aerial Training Group) No. 14 'Escorpiones' (Scorpions) at Mariscal Sucre, based in Mariscal Sucre, Maracay, it has a camouflage of Marine Corps Green (FS 34052), Light Green (FS 34227) and Light Tan (FS 30450). Undersides are Surface Grey (FS 36559).

Another EMB-312 of Grupo de Entrenamiento Aéreo No. 14 at Mariscal Sucre, FAV 4000 is shown as it appeared in 2013. It wears training colours of flat Silver (FS 17178), a red stripe in middle of the fuselage and the 'E' meaning Escuela (School). It is also armed for COIN or weapons training missions, with the LAU-32 pod with seven 2.75in (70mm) rockets and FN Herstal Heavy Machine Pod (HMP) with a 12.7mm (0.5in) M3P machine gun.

One of the most exuberant commemorative schemes belongs to EMB-312 FAV 5050 of Grupo de Entrenamiento Aéreo No. 14 at Mariscal Sucre, celebrating 100,000 flight hours. It also celebrates 27 years of the EMB-312 in Venezuela (1986 to 2013) and 92 years of Grupo de Entrenamiento Aéreo No. 14. Colour is overall gloss Black (FS 27040) with DayGlo nose (FS 38915) and black spinner.

APPENDIX III: ORBAT

Embraer EMB-312 Tucano and Short S-312 Tucano units

ANGOLA | Força Aérea Nacional de Angola (FANA, Angola Air Force)

Comando Operacional de Aviação de Transporte

Esquadra de Reconhecimento e Ataque, later renamed to Esquadra 401
Luanda
1998 – today

Escola Militar de Formação Aeronáutica do Lobito
Benguela
2002 – 2013

ARGENTINA | Fuerza Aérea Argentina (FAA, Argentine Air Force)

Escuela de Aviación Militar

Grupo Aéreo Escuela

Escuadrón II Tucano
Córdoba
1987 – today

BRAZIL | Força Aérea Brasileira (FAB, Brazilian Air Force)

Academia da Força Aérea
Pirassununga, São Paulo

1° Esquadrão de Instrução Aérea
1983 – today

Esquadrão de Demonstração Aérea (Esquadrilha da Fumaça)
1983 – 2013

7° Esquadrão de Emprego Tático e Adestramento (ETA)
Base Aérea de Manaus, Manaus
1993 – 1995

2° Esquadrilha de Ligação e Observação (ELO)
Base Aeronaval de São Pedro da Aldeia, Rio de Janeiro (1986 – 1995), Base Aérea de Santa Cruz, Rio de Janeiro (1995 – 2004)
1986 – 2004

1°/3° GAV 'Esquadrão Escorpião'
Base Aérea de Boa Vista, Roraima
1995 – 2008

1°/5° GAV 'Esquadrão Rumba'
Base Aérea de Natal, Rio Grande do Norte
1987 – 2002

1°/14° GAV 'Esquadrão Pampa'
Base Aérea de Canoas, Rio Grande do Sul
1990 – 2009

2°/3° GAV 'Esquadrão Grifo'
Base Aérea de Porto Velho, Rondônia
1995 – 2008

3°/3° GAV 'Esquadrão Flecha'
Base Aérea de Campo Grande, Mato Grosso do Sul
2003 – 2008

1° Grupo de Aviação de Caça (GAVCA)
1990 – 2009

1°/1° GAVCA 'Esquadrão Senta a Púa!'
1990 – 2009

2°/1° GAVCA 'Esquadrão Rompe Mato'
Base Aérea de Santa Cruz, Rio de Janeiro
1990 – 2009

1° Grupo de Defesa Aérea (GDA)
Base Aérea de Anápolis, Goiás
1990 – 2009

Instituto de Pesquisas e Ensaios em Voo
São José dos Campos, São Paulo
1983 – 2012

Parque de Material Aeronáutico de Lagoa Santa (PAMA LS)
Lagoa Santa, Minas Gerais
1982 – today

COLOMBIA | Fuerza Aérea Colombiana (FAC, Colombian Air Force)

Escuela Militar de Aviación
'Marco Fidel Suarez'
Vale del Cauca
1992 – 1993

Grupo de Combate 21

Escuadrón de Combate 212
'Tucanos'
Base Aérea de Apiay
Villavicencio, Meta
1993 – today

Grupo de Combate 61

Escuadrón de Combate 611
Base Aérea de Tres Esquinas,
Solano, Caquetá
2000 – today

EGYPT | Al Quwwat al Jawwiya al Misriya (Egyptian Air Force)

Egyptian Air Force Academy
Bilbeis Air Force Base, Bilbeis

101st Air Brigade
1985 – 2005

Egyptian Air Force Academy
Bilbeis Air Force Base, Bilbeis

201st Air Brigade
2001 – today

FRANCE | Armée de l'Air (AdlA, French Air Force)

École de l'Air

Centre de Formation Aéronautique
Militaire Initial 05/312
Salon de Provence Air Base, Marseille
1995 - 2009

HONDURAS | Fuerza Aérea Hondureña (FAH, Honduran Air Force)

Academia Militar
de Aviación

1° Escuadrón de Entreinamiento
Base Aérea Coronel José Enrique Soto Cano,
Palmerola
1984 - today

IRAN | Islamic Revolutionary Guard of Corps Air & Space Force, (IRGC ASF)

Aviation Training Centre
Badr Air Base, Isfahan
1989 - today

IRAQ | Al Quwwat Al Jawwiya Al Iraqiya (Iraq Air Force)

Iraq Air Force Academy
Tikrit
1984 – 2003

KENYA | Kenya Air Force

Kenya Air Force Flying Training School
Moi Air Base, Nairobi
1990 – today

KUWAIT | Al Quwwat al Jawwiya al Kuwaitiya (KAF, Kuwait Air Force)

19th Flight Training Squadron
Ali Al Salem Air Base
1995 – today

MAURITANIA | Force Aérienne Islamique de Mauritanie (FAIM, Air Force of Mauritania)

École Militaire de l'Air
Base Aérienne 210 Atar, Atar
2010 – today

PARAGUAY | Fuerza Aérea Paraguaya (FAP, Paraguaian Air Force)

Grupo Aerotáctico da FAP

Tercer Escuadrón de Caza 'Moros'
Base Aérea Ñu Guazú, Luque
1987 – today

PERU | Fuerza Aérea del Peru (FAPe, Peruvian Air Force)

Grupo Aéreo No. 51
Base Aérea de Las Palmas (1987–1991)
Base Aérea de Pisco (1991–today)

Escuadrón Aéreo No. 512
1987 – today

Escuadrón Aéreo No. 514
1991 – today

UNITED KINGDOM | Royal Air Force (RAF)

1 Flying Training School (FTS)
Linton-on-Ouse
1992 – today

72 (Reserve) Squadron
Linton-on-Ouse
2000 – today

76 (Reserve) Squadron
Linton-on-Ouse
2007-2011

207 (Reserve) Squadron
Linton-on-Ouse
2000 – today

3 Flying Training School (FTS)
Cranwell
1990 – 1993

6 Flying Training School (FTS)
Finningley
1991 – 1996

7 Flying Training School (FTS)
Church Fenton
1989 – 1992

Central Flying School
Linton-on-Ouse
1989 – today

Empire Test Pilots' School (ETPS)
Boscombe Down
1993 – today

VENEZUELA | Aviación Militar Bolivariana (AMB, Venezuelan Air Force)

Grupo de Entrenamiento Aéreo (GEA) No. 14 'Escorpiones'
Base Aérea Mariscal Sucre, Maracay

Escuadrón de Entrenamiento Avanzado No. 142
1986 – today

Escuadrón de Entrenamiento Táctico No. 143
2014 – today

Grupo Aéreo de Operaciones Especiales No. 15 'Zorros"

Escuadrón No. 152 'Avispones'
Base Aérea Rafael Urdaneta, Maracaibo
1987 – 1991
2001 – 2010

Grupo de Entrenamiento Táctico No. 13

Escuadrón No. 131 'Leones'
Base Aérea Luis Del Valle, Barcelona
1991 – 2001

APPENDIX IV

Angola | Força Aérea Nacional de Angola (FANA, Angola Air Force)

	Registration	c/n	Remarks
1	R-450	312429	new from Embraer
2	R-451	312430	new from Embraer
3	R-452	312431	new from Embraer
4	R-453	312432	new from Embraer
5	R-454	312433	new from Embraer; w/o, date unknown.
6	R-455	312434	new from Embraer
7	R-456	312053	purchased from Embraer, former Embraer demonstrator PT-ZTT
8	R-457	312149	purchased from Embraer, former Embraer demonstrator PT-ZTK
9	R-458	312283	purchased from Fuerza Aérea del Peru, former FAP 425
10	R-459	312285	purchased form Fuerza Aérea del Peru, former FAP 427
11	R-460	312298	purchased form Fuerza Aérea del Peru, former FAP 436
12	R-461	312304	purchased from Fuerza Aérea del Peru, former FAP 437
13	R-462	312314	purchased from Fuerza Aérea del Peru, former FAP 444; w/o 27 May 2008, crashed at Kissama during a training mission from Lubango to Luanda; both pilots, Major Cristóvão Saturnino Jamba and Lt Desejo Mendes Baltazar, were killed.
14	R-463	312403	purchased from Fuerza Aérea del Peru, former FAP 448

Argentina | Fuerza Aérea Argentina (FAA, Argentine Air Force)

	Registration	c/n	Remarks
1	E-101	312190	former Força Aérea Brasileira T-27 1417
2	E-102	312154	former Força Aérea Brasileira T-27 1399
3	E-103	312197	former Força Aérea Brasileira T-27 1420
4	E-104	312159	former Força Aérea Brasileira T-27 1401
5	E-105	312162	former Força Aérea Brasileira T-27 1403
6	E-106	312163	former Força Aérea Brasileira T-27 1404
7	E-107	312168	former Força Aérea Brasileira T-27 1406
8	E-108	312169	former Força Aérea Brasileira T-27 1407
9	E-109	312170	former Força Aérea Brasileira T-27 1408
10	E-110	312171	former Força Aérea Brasileira T-27 1409
11	E-111	312175	former Força Aérea Brasileira T-27 1410
12	E-112	312148	former Força Aérea Brasileira T-27 1397
13	E-113	312178	former Força Aérea Brasileira T-27 1412; w/o 29 June 2015.
14	E-114	312182	former Força Aérea Brasileira T-27 1413
15	E-115	312196	former Força Aérea Brasileira T-27 1419; w/o 30 May 1990, crashed near Las Varillas.
16	E-116	312322	new from Embraer
17	E-117	312329	new from Embraer

18	E-118	312331	new from Embraer; w/o 07 August 1988, crashed at Cordoba.
19	E-119	312332	new from Embraer
20	E-120	312336	new from Embraer
21	E-121	312337	new from Embraer
22	E-122	312341	new from Embraer
23	E-123	312342	new from Embraer
24	E-124	312357	new from Embraer
25	E-125	312358	new from Embraer
26	E-126	312360	new from Embraer
27	E-127	312361	new from Embraer; w/o 20 December 1988, crashed near Las Lagunillas, Cordoba.
28	E-128	312366	new from Embraer
29	E-129	312367	new from Embraer
30	E-130	312372	new from Embraer
31	E-131	312023	loaned from Força Aérea Brasileira 2009-2010, former T-27 1319
32	E-132	312058	loaned from Força Aérea Brasileira 2009-2010, former T-27 1344
33	E-133	312083	loaned from Força Aérea Brasileira 2009-2010, former T-27 1369
34	E-134	312096	loaned from Força Aérea Brasileira 2009-2010, former T-27 1380
35	E-135	312383	loaned from Força Aérea Brasileira 2009-2010, former T-27 1438

Brazil | Força Aérea Brasileira (FAB, Brazilian Air Force)

	Registration	c/n	Remarks
1	YT-27 1300	312001	First prototype; w/o 09 April 1984.
2	YT-27 1301	312002	Second prototype; w/o 09 August 1982.
3	T-27 1302	312005	
4	T-27 1303	312006	
5	T-27 1304	312008	w/o 21 October 1988.
6	T-27 1305	312009	w/o 02 November 1989.
7	T-27 1306	312010	w/o 28 October 1986.
8	T-27 1307	312011	w/o 02 April 2010.
9	T-27 1308	312012	
10	T-27 1309	312013	w/o 13 April 1986.
11	T-27 1310	312014	
12	T-27 1311	312015	w/o 22 September 1983.
13	T-27 1312	312016	
14	T-27 1313	312017	
15	T-27 1314	312018	
16	T-27 1315	312019	
17	T-27 1316	312020	w/o 27 April 1998.
18	T-27 1317	312039	w/o 29 May 1985.
19	T-27 1318	312022	
20	T-27 1319	312023	2009-2010 loaned to Fuerza Aérea Argentina as E-131; sold to Fuerza Aérea Paraguaya as FAP 1057 in 2010.
21	T-27 1320	312024	w/o 16 November 1996.
22	T-27 1321	312025	
23	T-27 1322	312026	
24	T-27 1323	312027	w/o 01 December 1987.
25	T-27 1324	312028	w/o 31 October 2007.
26	T-27 1325	312029	w/o 01 May 1989.
27	T-27 1326	312030	
28	T-27 1327	312031	
29	T-27 1328	312032	w/o 22 September 1993.
30	T-27 1329	312033	
31	T-27 1330	312034	

32	T-27 1331	312035	
33	T-27 1332	312036	
34	T-27 1333	312037	
35	T-27 1334	312038	
36	T-27 1335	312040	
37	T-27 1336	312041	
38	T-27 1337	312042	
39	T-27 1338	312043	removed from service, preserved at EPCAR.
40	T-27 1339	312044	
41	T-27 1340	312054	
42	T-27 1341	312055	
43	T-27 1342	312056	w/o 01 April 2003.
44	T-27 1343	312057	w/o 05 April 1987.
45	T-27 1344	312058	2009-2010 loaned to Fuerza Aérea Argentina as E-132; sold to Fuerza Aérea Paraguaya as FAP 1058 in 2010.
46	T-27 1345	312059	
47	T-27 1346	312060	
48	T-27 1347	312061	sold to Fuerza Aérea Paraguaya as FAP 1059 in 2010.
49	T-27 1348	312062	
50	T-27 1349	312063	
51	T-27 1350	312064	w/o 21 June 1997.
52	T-27 1351	312065	
53	T-27 1352	312066	
54	T-27 1353	312067	w/o 26 November 1992.
55	T-27 1354	312068	preserved at EDA hangar in Pirassununga, hanging upside down from hangar ceiling, in previous, red EDA colours.
56	T-27 1355	312069	
57	T-27 1356	312070	
58	T-27 1357	312071	
59	T-27 1358	312072	
60	T-27 1359	312073	
61	T-27 1360	312074	
62	T-27 1361	312075	in '30 anos' colours
63	T-27 1362	312076	w/o 13 Feburary 2004.
64	T-27 1363	312077	
65	T-27 1364	312078	
66	T-27 1365	312079	
67	T-27 1366	312080	w/o 24 March 2008.
68	T-27 1367	312081	preserved at EDA hangar in Pirassununga, hanging upside down from hangar ceiling; in EDA colours
69	T-27 1368	312082	
70	T-27 1369	312083	2009-2010 loaned to Fuerza Aérea Argentina as E-133
71	T-27 1370	312084	w/o 27 April 1994.
72	T-27 1371	312086	
73	T-27 1372	312087	
74	T 27 1373	312088	
75	T-27 1374	312089	preserved in a public square in Pirassununga.
76	T-27 1375	312091	
77	T-27 1376	312092	
78	T-27 1377	312093	
79	T-27 1378	312094	
80	T-27 1379	312095	
81	T-27 1380	312096	2009-2010 loaned to Fuerza Aérea Argentina as E-134
82	T-27 1381	312097	
83	T-27 1382	312100	

84	T-27 1383	312101	
85	T-27 1384	312102	
86	T-27 1385	312103	
87	T-27 1386	312104	
88	T-27 1387	312120	
89	T-27 1388	312121	w/o 02 October 1991.
90	T-27 1389	312137	
91	T-27 1390	312138	w/o 02 June 1994.
92	T-27 1391	312142	
93	T-27 1392	312143	
94	T-27 1393	312144	
95	T-27 1394	312145	
96	T-27 1395	312146	
97	T-27 1396	312147	
98	T-27 1397	312148	returned to Embraer in 1987 and then sold to Fuerza Aérea Argentina as E-112.
99	T-27 1398	312150	
100	T-27 1399	312154	returned to Embraer in 1987 and then sold to Fuerza Aérea Argentina as E-102.
101	T-27 1400	312155	
102	T-27 1401	312159	returned to Embraer in 1987 and then sold to Fuerza Aérea Argentina as E-104.
103	T-27 1402	312160	
104	T-27 1403	312162	returned to Embraer in 1987 and then sold to Fuerza Aérea Argentina as E-105.
105	T-27 1404	312163	returned to Embraer in 1987 and then sold to Fuerza Aérea Argentina as E-106.
106	T-27 1405	312167	
107	T-27 1406	312168	returned to Embraer in 1987 and then sold to Fuerza Aérea Argentina as E-107.
108	T-27 1407	312169	returned to Embraer in 1987 and then sold to Fuerza Aérea Argentina as E-108.
109	T-27 1408	312170	returned to Embraer in 1987 and then sold to Fuerza Aérea Argentina as E-109.
110	T-27 1409	312171	returned to Embraer in 1987 and then sold to Fuerza Aérea Argentina as E-110.
111	T-27 1410	312175	returned to Embraer in 1987 and then sold to Fuerza Aérea Argentina as E-111.
112	T-27 1411	312176	w/o 02 April 2005.
113	T-27 1412	312178	returned to Embraer in 1987 and then sold to Fuerza Aérea Argentina as E-113.
114	T-27 1413	312182	returned to Embraer in 1987 and then sold to Fuerza Aérea Argentina as E-114.
115	T-27 1414	312183	
116	T-27 1415	312184	w/o 09 November 2001.
117	T-27 1416	312189	
118	T-27 1417	312190	returned to Embraer in 1987 and then sold to Fuerza Aérea Argentina as E-101.
119	T-27 1418	312195	
120	T-27 1419	312196	returned to Embraer in 1987 and then sold to Fuerza Aérea Argentina as E-115.
121	T-27 1420	312197	returned to Embraer in 1987 and then sold to Fuerza Aérea Argentina as E-103.
122	T-27 1421	312369	w/o 02 June 1994.
123	T-27 1422	312374	
124	T-27 1423	312375	
125	T-27 1424	312378	
126	T-27 1425	312379	
127	T-27 1426	312380	
128	T-27 1427	312384	
129	T-27 1428	312385	
130	T-27 1429	312386	
131	T-27 1430	312389	
132	T-27 1431	312390	
133	T-27 1432	312391	w/o 21 December 1991.
134	T-27 1433	312393	w/o 13 February 2004.
135	T-27 1434	312394	
136	T-27 1435	312395	w/o 28 July 2016.
137	T-27 1436	312381	w/o 06 February 1991.

138	T-27 1437	312382	
139	T-27 1438	312383	2009-2010 loaned to Fuerza Aérea Argentina as E-135
140	T-27 1439	312387	
141	T-27 1440	312388	
142	T-27 1441	312392	
143	T-27 1442	312396	
144	T-27 1443	312397	
145	T-27 1444	312398	
146	T-27 1445	312399	w/o 01 May 1995.
147	T-27 1446	312424	
148	T-27 1447	312425	
149	T-27 1448	312426	
150	T-27 1449	312427	
151	T-27 1450	312428	

Colombia | Fuerza Aérea Colombiana (FAC, Colombian Air Force)

	Registration	c/n	Remarks
1	FAC 2250	312410	named 'Bull'
2	FAC 2251	312411	
3	FAC 2252	312412	named 'Baco'; still in original colours
4	FAC 2253	312413	named 'Spider'; grey colours
5	FAC 2254	312414	
6	FAC 2255	312415	
7	FAC 2256	312416	in 80.000 hrs marking plus shark mouth
8	FAC 2257	312417	named 'Espia'; grey colours
9	FAC 2258	312418	
10	FAC 2259	312419	
11	FAC 2260	312420	
12	FAC 2261	312421	named 'Llanera'; grey colours
13	FAC 2262	312422	named 'Agatha'; grey colours
14	FAC 2263	312423	named 'Daga'; grey colours

Egypt | Al Quwwat al Jawwiya al Misriya (Egyptian Air Force)

	Registration	c/n	Remarks
1	0501	312233	assembled at Kader Factory, Egypt
2	0502	312234	assembled at Kader Factory, Egypt
3	0503	312235	assembled by Kader Factory, Egypt; w/o 04 May 1990.
4	0504	312236	assembled at Kader Factory, Egypt
5	0505	312242	assembled at Kader Factory, Egypt
6	0506	312243	assembled at Kader Factory, Egypt
7	0507	312244	assembled at Kader Factory, Egypt
8	0508	312245	assembled at Kader Factory, Egypt
9	0509	312251	assembled at Kader Factory, Egypt
10	0510	312252	assembled at Kader Factory, Egypt
11	0511	312253	assembled at Kader Factory, Egypt
12	0512	312254	assembled at Kader Factory, Egypt
13	0513	312260	assembled at Kader Factory, Egypt
14	0514	312261	assembled at Kader Factory, Egypt
15	0515	312262	assembled at Kader Factory, Egypt

16	0516	312263	assembled at Kader Factory, Egypt
17	0517	312269	assembled at Kader Factory, Egypt
18	0518	312270	assembled at Kader Factory, Egypt
19	0519	312271	assembled at Kader Factory, Egypt
20	0520	312272	assembled at Kader Factory, Egypt
21	0521	312279	assembled at Kader Factory, Egypt
22	0522	312280	assembled at Kader Factory, Egypt
23	0523	312281	assembled at Kader Factory, Egypt
24	0524	312282	assembled at Kader Factory, Egypt
25	0525	312289	assembled at Kader Factory, Egypt
26	0526	312290	assembled at Kader Factory, Egypt
27	0527	312291	assembled at Kader Factory, Egypt
28	0528	312292	assembled at Kader Factory, Egypt
29	0529	312299	assembled at Kader Factory, Egypt
30	0530	312300	assembled at Kader Factory, Egypt
31	0531	312301	assembled at Kader Factory, Egypt
32	0532	312302	assembled at Kader Factory, Egypt
33	0533	312309	assembled at Kader Factory, Egypt
34	0534	312310	assembled at Kader Factory, Egypt
35	0535	312311	assembled at Kader Factory, Egypt
36	0536	312312	assembled at Kader Factory, Egypt
37	0537	312325	assembled at Kader Factory, Egypt
38	0538	312326	assembled at Kader Factory, Egypt
39	0539	312327	assembled at Kader Factory, Egypt
40	0540	312328	assembled at Kader Factory, Egypt
41	0541	312440	assembled at Kader Factory, Egypt
42	0542	312441	assembled at Kader Factory, Egypt
43	0543	312442	assembled at Kader Factory, Egypt
44	0544	312443	assembled at Kader Factory, Egypt
45	0545	312444	assembled at Kader Factory, Egypt
46	0546	312445	assembled at Kader Factory, Egypt
47	0547	312446	assembled at Kader Factory, Egypt
48	0548	312447	assembled at Kader Factory, Egypt
49	0549	312448	assembled at Kader Factory, Egypt
50	0550	312449	assembled at Kader Factory, Egypt
51	0551	312450	assembled at Kader Factory, Egypt
52	0552	312451	assembled at Kader Factory, Egypt
53	0553	312452	assembled at Kader Factory, Egypt
54	0554	312453	assembled at Kader Factory, Egypt

France I Armée de l'Air (AdlA, French Air Force)

	Registration	c/n	Remarks
1	312-UW 438	312438	stored
2	312-UY 439	312439	stored
3	312-JA 456	312456	stored
4	312-JB 457	312457	stored
5	312-JC 458	312458	stored
6	312-JD 459	312459	stored
7	312-JE 460	312460	stored dismantled in restoration hangar at Musée Européen de l'Aviation de Chasse, Montelimar, France.
8	312-JF 461	312461	sold in US as N461EY.
9	312-JG 462	312462	sold in US as N462MT.

10	312-JH 463	312463	sold to Mauritania Air Force as 5T-MAI.
11	312-JI 464	312464	sold in US as N464MT.
12	312-JJ 465	312465	w/o 12 February 1997.
13	312-JK 466	312466	stored
14	312-JL 467	312467	stored
15	312-JM 468	312468	stored
16	312-JN 469	312469	stored
17	312-JO 470	312470	stored
18	312-JP 471	312471	stored
19	312-JQ 472	312472	stored
20	312-JR 473	312473	sold in US as N473MT.
21	312-JS 474	312474	sold in US as N474MT.
22	312-JT 475	312475	stored
23	312-JU 477	312477	preserved and displayed in special colours at Rochefort-Saint-Agnant airport (BA 721, LFDN).
24	312-JV 478	312478	stored
25	312-JX 479	312479	stored
26	312-JY 480	312480	sold in US as N780MT.
27	312-JZ 481	312481	stored
28	312-UA 482	312482	w/o, date not known.
29	312-UB 483	312483	stored
30	312-UC 484	312484	sold in US as N484MT.
31	312-UD 485	312485	stored
32	312-EU 486	312486	used as ground trainer
33	312-UF 487	312487	sold to Mauritania Air Force as 5T-MAO.
34	312-UG 488	312488	sold to Mauritania Air Force as 5T-MAP.
35	312-UH 489	312489	stored
36	312-UI 490	312490	used as ground trainer
37	312-UJ 491	312491	stored
38	312-UK 492	312492	sold in US as N492MT.
39	312-UL 493	312493	used as ground trainer
40	312-UM 494	312494	used as ground trainer
41	312-UM 495	312495	stored
42	312-UO 496	312496	restored and preserved at Museé de l'Aire, Le Bourget.
43	312-UP 497	312497	stored
44	312-UQ 498	312498	preserved at CANOPEE Museum at Base Aérienne 279 'Lieutenant Beau', Chateaudun in France.
45	312-UR	312499	sold to Mauritania Air Force as 5T-MAT.
46	312-US	312500	used as ground trainer
47	312-UT	312501	sold in US as N501MT.
48	312-UU	312502	w/o 19 September 2003.
49	312-UV	312503	stored
50	312-UX	312504	preserved at Musee Europeen de l Aviation de Chasse, Montélimar, France.

Honduras | Fuerza Aérea Hondureña (FAH, Honduran Air Force)

	Registration	c/n	Remarks
1	FAH 0250	312045	
2	FAH 0251	312046	
3	FAH 0252	312047	
4	FAH 0253	312048	w/o 03 November 2004.
5	FAH 0254	312049	
6	FAH 0255	312050	w/o 05 October 2000.
7	FAH 0256	312051	
8	FAH 0257	312052	
9	FAH 0258	312198	emergency landing in a field near Comayagua, 3 March 2011

10	FAH 0259	312199	
11	FAH 0260	312205	
12	FAH 0261	312206	w/o 09 December 1987.

Iran | Islamic Revolutionary Guard of Corps Air & Space Force

	Registration	c/n	Remarks
1	15-2401	312340	
2	15-2402	312343	
3	15-2403	312344	
4	15-2404	312356	
5	15-2405	312359	
6	15-2406	312362	
7	15-2407	312363	
8	15-2408	312364	
9	15-2409	312365	
10	15-2410	312368	
11	15-2411	312370	
12	15-2413	312371	
13	15-2413	312373	
14	15-2414	312376	
15	15-2415	312377	

Iraq | Al Quwwat Al Jawwiya Al Iraqiya (Iraq Air Force)

	Registration	c/n	Remarks
1	6700	312085	
2	6701	312090	
3	6702	312098	
4	6703	312099	
5	6710	312105	assembled at Kader Factory, Egypt
6	6711	312106	assembled at Kader Factory, Egypt
7	6712	312107	assembled at Kader Factory, Egypt
8	6704	312108	
9	6705	312109	
10	6713	312110	assembled at Kader Factory, Egypt
11	6714	312111	assembled at Kader Factory, Egypt
12	6715	312112	assembled at Kader Factory, Egypt
13	6706	312113	
14	6707	312114	
15	6716	312115	assembled at Kader Factory, Egypt
16	6717	312116	assembled at Kader Factory, Egypt
17	6718	312117	assembled at Kader Factory, Egypt
18	6708	312118	
19	6709	312119	
20	6719	312122	assembled at Kader Factory, Egypt
21	6720	312123	assembled at Kader Factory, Egypt
22	6721	312124	assembled at Kader Factory, Egypt
23	6722	312125	assembled at Kader Factory, Egypt
24	6723	312126	assembled at Kader Factory, Egypt
25	6724	312127	assembled at Kader Factory, Egypt
26	6725	312128	assembled at Kader Factory, Egypt
27	6726	312129	assembled at Kader Factory, Egypt

28	6727	312130	assembled at Kader Factory, Egypt
29	6728	312131	assembled at Kader Factory, Egypt
30	6729	312132	assembled at Kader Factory, Egypt
31	6730	312133	assembled at Kader Factory, Egypt
32	6731	312134	assembled at Kader Factory, Egypt
33	6732	312135	assembled at Kader Factory, Egypt
34	6733	312136	assembled at Kader Factory, Egypt
35	6734	312139	assembled at Kader Factory, Egypt
36	6735	312140	assembled at Kader Factory, Egypt
37	6736	312141	assembled at Kader Factory, Egypt
38	6737	312151	assembled at Kader Factory, Egypt
39	6738	312152	assembled at Kader Factory, Egypt
40	6739	312153	assembled at Kader Factory, Egypt
41	6740	312156	assembled at Kader Factory, Egypt
42	6741	312157	assembled at Kader Factory, Egypt
43	6742	312158	assembled at Kader Factory, Egypt
44	6743	312164	assembled at Kader Factory, Egypt
45	6744	312165	assembled at Kader Factory, Egypt
46	6745	312166	assembled at Kader Factory, Egypt
47	6746	312172	assembled at Kader Factory, Egypt
48	6747	312173	assembled at Kader Factory, Egypt
49	6748	312174	assembled at Kader Factory, Egypt
50	6749	312179	assembled at Kader Factory, Egypt
51	6750	312180	assembled at Kader Factory, Egypt
52	6751	312181	assembled at Kader Factory, Egypt
53	6752	312185	assembled at Kader Factory, Egypt
54	6753	312186	assembled at Kader Factory, Egypt
55	6754	312187	assembled at Kader Factory, Egypt
56	6755	312188	assembled at Kader Factory, Egypt
57	6756	312191	assembled at Kader Factory, Egypt
58	6757	312192	assembled at Kader Factory, Egypt
59	6758	312193	assembled at Kader Factory, Egypt
60	6759	312194	assembled at Kader Factory, Egypt
61	6760	312201	assembled at Kader Factory, Egypt
62	6761	312202	assembled at Kader Factory, Egypt
63	6762	312203	assembled at Kader Factory, Egypt
64	6763	312204	assembled at Kader Factory, Egypt
65	6764	312210	assembled at Kader Factory, Egypt
66	6765	312211	assembled at Kader Factory, Egypt
67	6766	312212	assembled at Kader Factory, Egypt
68	6767	312213	assembled at Kader Factory, Egypt
69	6768	312220	assembled at Kader Factory, Egypt
70	6769	312221	assembled at Kader Factory, Egypt
71	6770	312222	assembled at Kader Factory, Egypt
72	6771	312223	assembled at Kader Factory, Egypt
73	6772	312229	assembled at Kader Factory, Egypt
74	6773	312230	assembled at Kader Factory, Egypt
75	6774	312231	assembled at Kader Factory, Egypt
76	6775	312232	assembled at Kader Factory, Egypt
77	6776	312318	assembled at Kader Factory, Egypt
78	6777	312319	assembled at Kader Factory, Egypt
79	6778	312320	assembled at Kader Factory, Egypt
80	6779	312321	assembled at Kader Factory, Egypt

Kenya | Kenya Air Force

	Registration	c/n	Remarks
1	811	S36-E2	Shorts Tucano T.51
2	812	S42-E3	Shorts Tucano T.51
3	813	S47-E4	Shorts Tucano T.51
4	814	S49-E5	Shorts Tucano T.51
5	815	S55-E6	Shorts Tucano T.51
6	816	S57-E7	Shorts Tucano T.51
7	817	S59-E8	Shorts Tucano T.51
8	818	S65-E13	Shorts Tucano T.51
9	819	S66-E14	Shorts Tucano T.51
10	820	S72-E19	Shorts Tucano T.51
11	821	S82-E28	Shorts Tucano T.51
12	822	S83-E29	Shorts Tucano T.51

Kuwait | Al Quwwat al Jawwiya al Kuwaitiya (KAF, Kuwait Air Force)

	Registration	c/n	Remarks
1	KAF 101	S60-E9	Shorts Tucano T.52; in storage since 2016
2	KAF 102	S61-E10	Shorts Tucano T.52; in storage since 2016
3	KAF 103	S62-E11	Shorts Tucano T.52; in storage since 2016
4	KAF 104	S63-E12	Shorts Tucano T.52; in storage since 2016
5	KAF 105	S67-E15	Shorts Tucano T.52; in storage since 2016
6	KAF 106	S68-E16	Shorts Tucano T.52; in storage since 2016
7	KAF 107	S69-E17	Shorts Tucano T.52; in storage since 2016
8	KAF 108	S71-E18	Shorts Tucano T.52; in storage since 2016
9	KAF 109	S74-E20	Shorts Tucano T.52; in storage since 2016
10	KAF 110	S75-E21	Shorts Tucano T.52; in storage since 2016
11	KAF 111	S76-E22	Shorts Tucano T.52; in storage since 2016
12	KAF 112	S77-E23	Shorts Tucano T.52; in storage since 2016
13	KAF 113	S80-E24	Shorts Tucano T.52; in storage since 2016
14	KAF 114	S81-E25	Shorts Tucano T.52; in storage since 2016
15	KAF 115	S89-E26	Shorts Tucano T.52; in storage since 2016
16	KAF 116	S90-E27	Shorts Tucano T.52; in storage since 2016

Mauretania | Force Aérienne Islamique de Mauritanie (FAIM, Air Force of Mauritania)

	Registration	c/n	Remarks
1	312463	5T-MAI	received from Armée del'Aire
2	312471	5T-MAA	received from Armée del'Aire
3	312487	5T-MAO	received from Armée del'Aire
4	312488	5T-MAP	received from Armée del'Aire
5	312499	5T-MAT	received from Armée del'Aire

Paraguay | Fuerza Aérea Paraguaya (FAP, Paraguaian Air Force)

	Registration	c/n	Remarks
1	FAP 1051	312330	
2	FAP 1052	312333	
3	FAP 1053	312334	

4	FAP 1054	312335	w/o 05 March 1995.
5	FAP 1055	312338	w/o 10 March 1998.
6	FAP 1056	312339	w/o 19 October 1989.
7	FAP 1057	312023	purchased from Força Aérea Brasileira, former T-27 1319.
8	FAP 1058	312058	purchased from Força Aérea Brasileira, former T-27 1344.
9	FAP 1059	312061	purchased from Força Aérea Brasileira, former T-27 1347.

Peru | Fuerza Aérea del Peru (FAPe, Peruvian Air Force)

	Registration	c/n	Remarks
1	421	312315	w/o 01 July 2003.
2	422	312316	
3	423	312317	
4	424	312323	
5	425	312283	sold to Força Aérea Nacional de Angola as R-458.
6	426	312284	w/o 24 May 1988, 3 aircraft collided during formation flight.
7	427	312285	sold to Força Aérea Nacional de Angola as R-459.
8	428	312286	w/o 24 May 1988, 3 aircraft collided during formation flight.
9	429	312287	
10	430	312288	w/o 24 May 1988, 3 aircraft collided during formation flight.
11	431	312293	
12	432	312294	
13	433	312295	w/o 10 December 1998.
14	434	312296	
15	435	312297	
16	436	312298	sold to Força Aérea Nacional de Angola as R-460.
17	437	312304	sold to Força Aérea Nacional de Angola as R-461.
18	438	312313	
19	439	312400	
20	440	312401	
21	443	312402	
22	444	312314	sold to Força Aérea Nacional de Angola as R-462.
23	448	312403	sold to Força Aérea Nacional de Angola as R-463.
24	449	312404	
25	450	312405	
26	451	312406	w/o 01 March 2001.
27	454	312407	
28	455	312408	
29	457	312409	
30	473	312324	

United Kingdom | Royal Air Force (RAF)

	Registration	c/n	Remarks
1	ZF135	S1-T1	
2	ZF136	S2-T2	exported to US as N885RS.
3	ZF137	S3-T3	
4	ZF138	S4-T4	exported to US as N884RS.
5	ZF139	S5-T5	
6	ZF140	S6-T6	
7	ZF141	S7-T7	sold to Everett Aero.
8	ZF142	S8-T8	
9	ZF143	S9-T9	

10	ZF144	S10-T10	
11	ZF145	S11-T11	
12	ZF160	S12-T12	sold to Everett Aero.
13	ZF161	S13-T13	exported to US.
14	ZF162	S14-T14	exported to US.
15	ZF163	S15-T15	stored at AM&SU (EGOS)
16	ZF164	S16-T16	exported to US as N192S.
17	ZF165	S17-T17	exported to US as N8093Z.
18	ZF166	S18-T18	stored at AM&SU (EGOS)
19	ZF167	S19-T19	fuselage perserved at Ulster Aviation Society Museum, Northern Ireland.
20	ZF168	S20-T20	exported to US as N838RS.
21	ZF169	S21-T21	
22	ZF170	S22-T22	
23	ZF171	S23-T23	
24	ZF172	S24-T24	
25	ZF200	S25-T25	exported to US as N822RS.
26	ZF201	S26-T26	exported to US.
27	ZF202	S27-T27	preserved at Linton-on-Ouse, UK.
28	ZF203	S28-T28	sold to Everett Aero.
29	ZF204	S29-T29	
30	ZF205	S30-T30	
31	ZF206	S31-T31	sold to Tucano Flyer LLC as N206PZ; w/o 22 June 2015.
32	ZF207	S32-T32	exported to US as N862RS.
33	ZF208	S34-T33	exported to US as N208PZ.
34	ZF209	S35-T34	sold to Everett Aero.
35	ZF210	S37-T35	
36	ZF211	S38-T36	stored at AM&SU (EGOS)
37	ZF212	S39-T37	sold to Everett Aero.
38	ZF238	S40-T38	exported to US as N833RS.
39	ZF239	S41-T39	
40	ZF240	S43-T40	
41	ZF241	S44-T41	exported to US as N835RS.
42	ZF242	S46-T43	sold to Everett Aero.
43	ZF243	S48-T44	
44	ZF244	S50-T45	
45	ZF245	S51-T46	sold to Everett Aero.
46	ZF263	S52-T47	stored at AM&SU (EGOS)
47	ZF264	S53-T48	
48	ZF265	S54-T49	exported to US as N1425S.
49	ZF266	S56-T50	exported to US as N846RS.
50	ZF267	S58-T51	exported to US as N267TC.
51	ZF268	S64-T52	stored at AM&SU (EGOS)
52	ZF269	S70-T53	
53	ZF270	S73-T54	w/o 13 May 1996.
54	ZF284	S78-T55	sold, departed by road.
55	ZF285	S79-T56	exported to US as N634CD.
56	ZF286	S84-T57	stored at AM&SU (EGOS)
57	ZF287	S85-T58	
58	ZF288	S86-T59	stored at AM&SU (EGOS)
59	ZF289	S87-T60	
60	ZF290	S88-T61	
61	ZF291	S91-T62	
62	ZF292	S92-T63	
63	ZF293	S93-T64	

64	ZF294	S94-T65	
65	ZF295	S95-T66	
66	ZF315	S96-T67	stored at AM&SU (EGOS)
67	ZF316	S97-T68	w/o 12 May 1992.
68	ZF317	S98-T69	
69	ZF318	S99-T70	stored at AM&SU (EGOS)
70	ZF319	S100-T71	stored at AM&SU (EGOS)
71	ZF320	S101-T72	exported to US as N883RS.
72	ZF338	S102-T73	
73	ZF339	S103-T74	
74	ZF340	S104-T75	sold to Everett Aero.
75	ZF341	S105-T76	
76	ZF342	S106-T77	
77	ZF343	S107-T78	
78	ZF344	S108-T79	w/o 12 March 2009.
79	ZF345	S109-T80	sold to Everett Aero.
80	ZF346	S110-T81	exported to US as N806RS.
81	ZF347	S111-T82	
82	ZF348	S112-T83	
83	ZF349	S113-T84	
84	ZF350	S114-T85	stored at AM&SU (EGOS)
85	ZF372	S115-T86	stored at AM&SU (EGOS)
86	ZF373	S116-T87	exported to US.
87	ZF374	S117-T88	
88	ZF375	S118-T89	exported to US (N845RS).
89	ZF376	S119-T90	stored at AM&SU (EGOS)
90	ZF377	S120-T91	
91	ZF378	S121-T92	
92	ZF379	S122-T93	
93	ZF380	S123-T94	stored at AM&SU (EGOS)
94	ZF405	S124-T95	stored at AM&SU (EGOS)
95	ZF406	S125-T96	
96	ZF407	S126-T97	
97	ZF408	S127-T98	stored at AM&SU (EGOS)
98	ZF409	S128-T99	
99	ZF410	S129-T100	sold to Everett Aero.
100	ZF411	S130-T101	exported to US as N411ZF.
101	ZF412	S131-T102	stored at AM&SU (EGOS)
102	ZF413	S132-T103	exported to US as N855RS.
103	ZF414	S133-T104	stored at AM&SU (EGOS)
104	ZF415	S134-T105	exported to US as N710RS.
105	ZF416	S135-T106	stored at AM&SU (EGOS)
106	ZF417	S136-T107	
107	ZF418	S137-T108	stored at AM&SU (EGOS)
108	ZF445	S138-T109	exported to US.
109	ZF446	S139-T110	stored at AM&SU (EGOS)
110	ZF447	S140-T111	sold to Everett Aero.
111	ZF448	S141-T112	
112	ZF449	S142-T113	stored at AM&SU (EGOS)
113	ZF450	S143-T114	exported to US as N8093S.
114	ZF483	S144-T115	sold to Everett Aero.
115	ZF484	S145-T116	exported to US as N373BR.
116	ZF485	S146-T117	
117	ZF486	S147-T118	stored at AM&SU (EGOS)

118	ZF487	S148-T119	stored at AM&SU (EGOS)
119	ZF488	S149-T120	stored at AM&SU (EGOS)
120	ZF489	S150-T121	
121	ZF490	S151-T122	stored at AM&SU (EGOS)
122	ZF491	S152-T123	
123	ZF492	S153-T124	stored at AM&SU (EGOS)
124	ZF510	S154-T125	
125	ZF511	S155-T126	
126	ZF512	S156-T127	
127	ZF513	S157-T128	stored at AM&SU (EGOS)
128	ZF514	S158-T129	sold to Everett Aero.
129	ZF515	S159-T130	stored at AM&SU (EGOS)
130	ZF516	S160-T131	sold to Everett Aero.

Venezuela I Aviación Militar Bolivariana (AMB, Venezuelan Air Force)

	Registration	c/n	Remarks
1	2820	312177	
2	3050	312207	
3	2370	312208	w/o 20 January 2004.
4	2460	312209	
5	0150	312215	
6	0040	312216	
7	0086	312217	
8	0100	312218	
9	0175	312219	
10	0214	312224	
11	1335	312225	
12	0203	312226	
13	0260	312228	
14	3105	312237	w/o 19 December 2001.
15	4000	312238	
16	4180	312239	
17	6888	312240	
18	1405	312241	
19	1792	312246	w/o 25 May 2001, collided in the air with FAV 9153.
20	5050	312247	
21	5613	312248	
22	9153	312249	w/o 25 May 2001, collided in the air with FAV 1792.
23	2280	312255	
24	6300	312256	w/o 12 December 1987.
25	7101	312257	
26	7321	312259	
27	8090	312264	w/o 22 March 2005.
28	8247	312265	
29	9244	312267	
30	5625	312268	
31	3131	312303	w/o 30 March 2008.

INDEX

Battles, Exercises and Operations

Personalities

Tucano Project and development

Units

COMMISSIONED SQUADRON LITHOGRAPHS
We Proudly Serve:
AIR SHOWS
AIR RACERS
LAW ENFORCEMENT
USNAVY
USAF
MARINES
US COAST GUARD
US ARMY
AVIATIONGRAPHIC.COM
Phone: +39.338.487.3171
Email: info@aviationgraphic.com
Address: Via Enrico Lai 35, 09128 Cagliari - ITALY
FULLY CUSTOMIZABLE LITHOGRAPHS WITH LOGO AND TEXT.
PERSONALIZE YOUR SQUADRON LITHOS WITH YOUR AIRCRAFT.
YOU CHOOSE THE CONTENTS, WE MAKE THE FINE ART PRINTS!
* FREE PREVIEWS AND HEAVY DISCOUNTS FOR BULK ORDERS *
www.aviationgraphic.com
We are on FACEBOOK
Join our community!